Interacting Bosons
in Nuclear Physics

ETTORE MAJORANA INTERNATIONAL SCIENCE SERIES

Series Editor:
Antonino Zichichi
European Physical Society
Geneva, Switzerland

(PHYSICAL SCIENCES)

Interacting Bosons in Nuclear Physics

Edited by

F. Iachello

University of Groningen
Groningen, The Netherlands
and
Yale University
New Haven, Connecticut

Plenum Press · New York and London

Library of Congress Cataloging in Publication Data

Symposium on Interacting Bosons in Nuclear Physics, 1st, Erice, Italy, 1978.
  Interacting bosons in nuclear physics.

  (Ettore Majorana international sciences series: Physical science; v. 1)
  Includes index.
  1. Bosons—Congresses. I. Iachello, F. II. Title. III. Series
QC793.5.B628S96 1978   539.7'21   79-13600

ISBN 978-1-4684-3523-8   ISBN 978-1-4684-3521-4 (eBook)
DOI 10.1007/978-1-4684-3521-4

Proceedings of the First Symposium on Interacting Bosons in Nuclear Physics, held in Erice, Sicily, June 6—9, 1978

© 1979 Plenum Press, New York
Softcover reprint of the hardcover 1st edition 1979
A Division of Plenum Publishing Corporation
227 West 17th Street, New York, N.Y. 10011

PREFACE

During the week of June 6-9, 1978, a group of 36 physicists
from 15 countries met in Erice, Sicily, for the first specialized
seminar on "Interacting Bosons in Nuclear Physics". The countries
represented were Argentina, Belgium, Denmark, Finland, France, the
Federal Republic of Germany, Israel, Italy, Japan, the Netherlands,
Poland, Sweden, the United Kingdom, the United States of America
and Yugoslavia. The Seminar was sponsored by the Italian Ministry
of Public Education (MPI), the Italian Ministry of Scientific and
Technological Research (MRST), the North Atlantic Treaty Organiza-
tion (NATO) and the Regional Sicilian Government (ERS).

The purpose of the Seminar was to discuss the present status of
the Interacting Boson Model both from the theoretical and experimental point of view. Some of the lectures presented in this book
summarize particular aspects of the model and are based on previously
published work (F. Iachello, R. F. Casten, Z. Sujkowski, L. Hasselgren, H. Emling, I. Talmi, T. Otsuka, J. McGrory, A.E.L. Dieperink
and A. Arima). Others are entirely new. In particular, the lectures of O. Scholten and A. Gelberg and V. Kaup present the first
extensive set of calculations based on the proton-neutron boson
model, while the lecture of J.N.Ginocchio describes a fermion model
with properties identical to those of the interacting boson model.
Also new are the lectures of D. R. Bes, R. A. Broglia and P. F.
Bortignon describing the relation of the interacting boson model
with Nuclear Field Theory, and of J. Mayer-ter-Vehn describing the
relation of the 0(6) limit of the interacting boson model with the
γ-unstable model of Wilets and Jean. Finally, the paper of V. Paar
takes a critical look at the relation between the interacting boson
model and the model of Janssen, Jolos and Dönau, while the paper of H.
Feshbach makes some general remarks on the difference between the
interacting boson model and interacting boson approximation.

Since the time of the Meeting in Erice, many new developments
have occurred on the subject, among others those related to collective states in odd-A nuclei. These developments are not included in
this book and hopefully will form the subject of a second Specialized Seminar.

I wish to thank all the participants and contributors who made the Seminar a very enjoyable one. I also wish to thank Professor A. Zichichi, Director of the Center for Scientific Culture "Ettore Majorana" who made it possible that this Seminar could take place in Erice, and Professor G. Preparata, Director of the Seminars, for his help and collaboration. A final thank you to all those who in Erice, Groningen and Yale have helped me on many occasions. Without them, the preparation of this book would not have been possible.

F. Iachello
February 1979
New Haven, Conn.

TABLE OF CONTENTS

CONTENTS   ix

PRESENT STATUS OF THE INTERACTING BOSON MODEL

F. Iachello

Kernfysisch Vernsneller Instituut

University of Groningen, Groningen, The Netherlands

1. INTRODUCTION

The purpose of this meeting is to discuss the scope and lim-itations of the interacting boson model[1] which has been intro-duced recently in order to describe collective low-lying states in medium and heavy-mass nuclei. In my talk, I will summarize the present status of the model referring to the following talks for more detailed and complete expositions. It seems appropri-ate to divide the talk in two parts in the first part, I will discuss the interacting boson model, as a tool to classify and describe collective states in nuclei; in the second part, I will discuss the interacting boson approximation in which one tries to derive the model as an approximation scheme to the shell model, thus providing a microscopic description of the collective states. While at present the first part may be, to a good extent, con-sidered concluded, in the sense that all possible solutions have been investigated, the second part is still in its infant stage, and it may require several years before a fully microscopic theory will be preserved.

2. THE MODEL

I begin with a short summary of the model. In its simplest form, we assume that an even-even nucleus consists of an inert core plus some valence particles. Furthermore, we assume that the valence particles, which are those outside the major closed shells at 50, 82, 126, tend to pair together in states with angular momentum L = 0 and 2, and we treat these pairs as bosons. The pairs with angular momentum L = 0 are called s-bosons, those

with angular momentum L = 2 are called d-bosons, and the total
number N of bosons in a given even-even nucleus is the sum of the
neutron, N_ν, and proton, N_π, pairs $N = N_\nu + N_\pi$. If more than half
of the shell is full, $N_{\pi(\nu)}$, is taken as the number of hole pairs.
Thus, for example, in

$$^{128}_{54}Xe_{74} \quad N = N_\pi + N_\nu = 2 + 4 = 6.$$

Introducing creation $(d^\dagger_\mu, s^\dagger)$ and annihilation (d_μ, s) operators,
the most general Hamiltonian, which includes one-boson terms and
boson-boson interactions, can be written as

$$H = \epsilon_s \, (s^\dagger.s) + \epsilon_d \, (d^\dagger.\tilde{d}) +$$

$$+ \sum_{L=0,2,4} \frac{1}{2} (2L+1)^{\frac{1}{2}} c_L \left[(d^\dagger \times d^\dagger)^{(L)} \times (\tilde{d} \times \tilde{d})^{(L)} \right]^{(0)}$$

$$+ \frac{1}{2^{\frac{1}{2}}} \tilde{v}_2 \left[(d^\dagger \times d^\dagger)^{(2)} \times (\tilde{d} \times s)^{(2)} + (d^\dagger \times s^\dagger)^{(2)} \times (\tilde{d} \times \tilde{d})^{(2)} \right]^{(0)}$$

$$+ \frac{1}{2} \tilde{v}_0 \left[(\tilde{d} \times d^\dagger)^{(0)} \times (s \times s)^{(0)} + (s^\dagger \times s^\dagger)^{(0)} \times (\tilde{d} \times \tilde{d})^{(0)} \right]^{(0)}$$

$$+ u_2 \left[(d^\dagger \times s^\dagger)^{(2)} \times (\tilde{d} \times s)^{(2)} \right]^{(0)} + \frac{1}{2} u_0 \left[(s^\dagger \times s^\dagger)^{(0)} \times (s \times s)^{(0)} \right]^{(0)}$$

$$(2.1)$$

Here the parameters $\epsilon_L (L=0,2)$, $c_L (L=0,2,4)$, $\tilde{v}_L (L=0,2)$, $u_L (L=0,2)$
describe the boson energies and interactions, $\tilde{d}_\mu = (-)^\mu d_{-\mu}$, and the
parenthesis denote angular momentum couplings. These parameters
depend explicitly on the boson number N and thus change from nucleus
to nucleus. One of the most important aspects of the interacting
boson model is the possibility to accommodate, as N changes, the
various situations encountered in nuclei.

In general, the eigenvalues and eigenstates can be found by
diagonalizing H in an appropriate basis. However, for some values
of N, it may happen that the Hamiltonian H takes on a form for which
analytic solutions can be found. In order to find these solutions,
we have taken advantage of the group structure of the problem. Since
the five ($\mu=0, \pm1, \pm2$) components of the d-boson and the single com-
ponent of the s-boson span a six-dimensional space, the group structure of the problem is that of U(6) (or SU(6) if we consider a fixed
boson number N). Analytic solutions can be found whenever the Hamiltonian H can be written in terms of invariants only of a complete
chain of subgroups of SU(6). There are three possible chains and
thus three possible analytic solutions, which for convenience, have

been labelled by the first subgroup of SU(6) which appears in the
chain: I) SU(5); II) SU(3) and III) 0(6). The properties of chains
I and II have been already extensively discussed[2,3]. Those of
chain III have been briefly described[4] and a long paper with a
complete description is now in preparation. To summarize these
three analytic solutions, I quote here their corresponding energy
formulas

I) SU(5)[2]

$$E([N],n_d,v,n_\Delta,L,M)=\varepsilon n_d+\alpha\tfrac{1}{2}n_d(n_d-1)+\beta(n_d-v)\ (n_d+v+3)+\gamma\left[L(L+1)-6n_d\right]$$
$$(2.2)$$

II) SU(3)[3]

$$E([N],(\lambda,\mu),K,L,M)=\ (\tfrac{3}{4}\kappa+\kappa')\ L(L+1)-\kappa\left[\lambda^2+\mu^2+\lambda\mu+3\ (\lambda+\mu)\right]\qquad(2.3)$$

III) 0(6)[4]

$$E([N],\sigma,\tau,v_\Delta,L,M)\ =A\ \tfrac{1}{4}\ (N-\sigma)\ (N+\sigma+4)\ +\ B\tau(\tau+3)+\ C\ L\ (L+1)\ .\quad(2.4)$$

Here the symbols in parenthesis after E denote the quantum numbers
which are needed to specify uniquely the states and the constants
$\varepsilon,\alpha,\beta,\gamma;\kappa,\kappa'$; A, B, C label boson energies and interactions in the
three limiting cases. They are a particular linear combination
of the parameters ε_L, c_L, v_L, u_L appearing in (2.1). The spectra
which correspond to these three limiting cases are shown in
Figs. 1, 2 and 3.

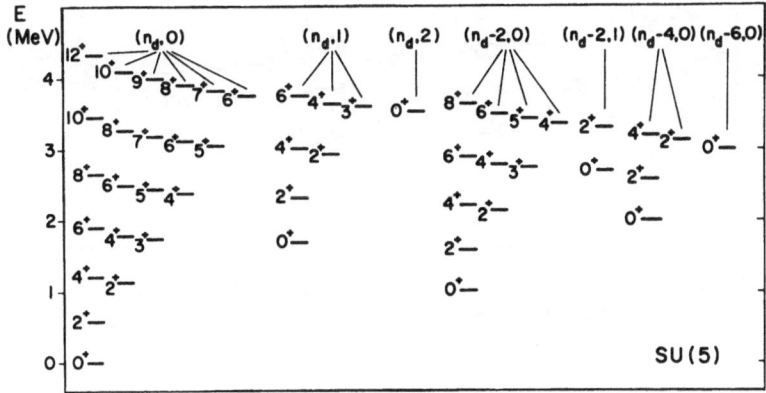

Fig. 1.   A typical spectrum with SU(5) symmetry and N=6.   In
          parenthesis are the values of v and n_Δ.

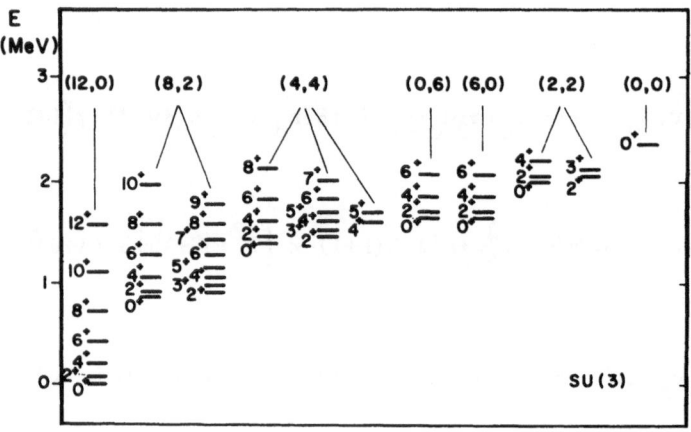

Fig. 2.   A typical spectrum with SU(3) symmetry and N=6.   In
          parenthesis are the values of λ and μ which label the
          SU(3) representations.

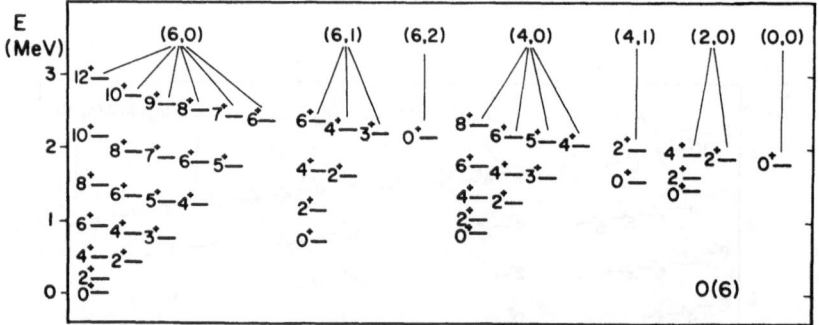

Fig. 3.   A typical spectrum with O(6) symmetry and N=6.   In
          parenthesis are the values of σ and ν_Δ.

Whenever the energy levels can be written in terms of the quantum numbers which uniquely specify the states one says that a <u>dynamical symmetry</u> occurs. One of the most interesting aspects of the interacting boson model is that of having suggested the occurrence of three types of dynamical symmetries. All three seem to be observed. Examples of the first two types were already given in Refs. 2 and 3. However, evidence for the third type of symmetry, 0(6), had not been given until recently[5] and I refer to Casten's talk for more details. Usually, examples of type I are found in nuclei at the beginning of shells, examples of type II in nuclei in the middle of the shells, and examples of type III in nuclei towards the end of the shells. For completeness, three examples are shown in Figs. 4, 5 and 6.

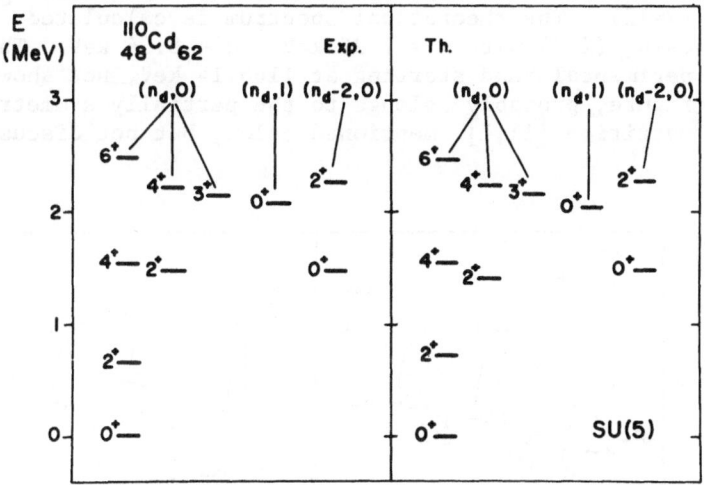

Fig. 4. An example[2] of a spectrum with SU(5) symmetry: $^{110}_{48}Cd_{62}$, (N=7). The theoretical energies are calculated using (2.2) with ϵ = 722 keV, α = 18 keV, β = 10.3 keV, γ = 10 keV.

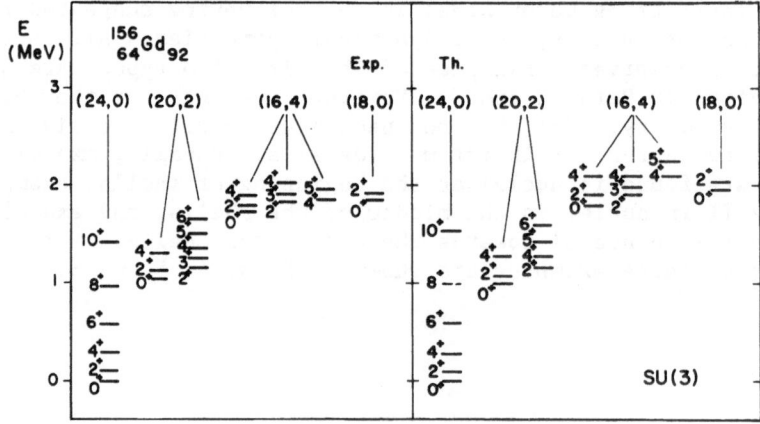

Fig. 5.   An example[3] of a spectrum with SU(3) symmetry: $^{156}_{64}$Gd$_{92}$,
          (N=12). The theoretical spectrum is calculated
          using (2.3) with κ = 7.25 keV, κ'= 8.56 keV. The ex-
          perimental band starting at 1168.14 keV, not shown in the
          figure, probably belongs to the partially symmetric
          partition [11,1] mentioned below, but not discussed here.

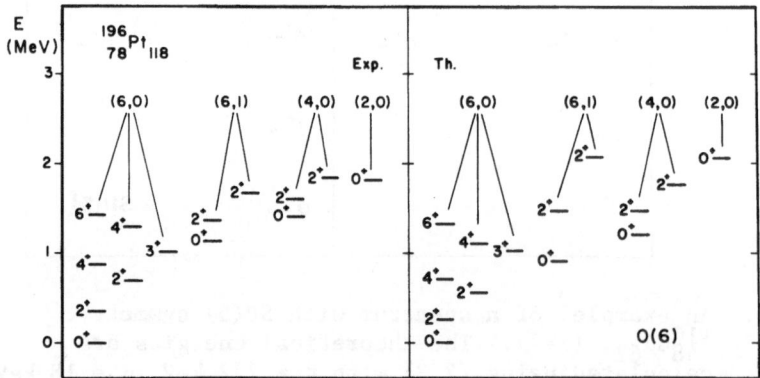

Fig. 6.   An example[5] of a spectrum with O(6) symmetry: $^{196}_{78}$Pt$_{118}$,
          (N=6). The theoretical spectrum is calculated
          using (2.4) with A=172 keV, B=50 keV, C=10 keV.

In addition to providing analytic solutions, the interacting boson model allows also to study the intermediate situations. For these one has to return to the full Hamiltonian (2.1) and diagonalize it. A computer program, called PHINT, has been written for this purpose and is available on request. I would like to divide the intermediate situations into three classes: A) between I and II; B) between II and III and C) between III and I. The particular features of class A have been analyzed in Ref. 6. Some typical properties are shown in Figs. 7 and 8.

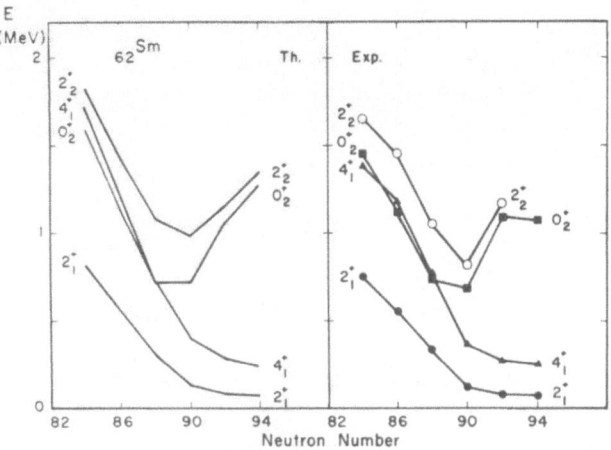

Fig. 7. Typical features of the transitional class A^6. Energies.

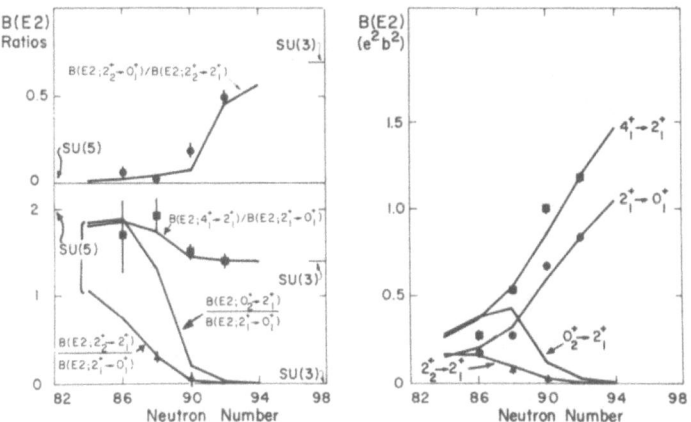

Fig. 8. Typical features of the transitional class A^6. Electromagnetic transition rates.

Note in these figures the drop and rise of the states 0_2^+ and 2_2^+ and the behavior of the branching ratio $R=B(E2;2_2^+\rightarrow0_1^+)/B(E2;\ 2_2^+\rightarrow2_1^+)$ which changes from the value zero appropriate to the coupling scheme I to the value 0.7 appropriate to the coupling scheme II. Class B has been analyzed in Ref. 7 and Casten will discuss in his talk its properties in detail. Some typical features of this transitional class of nuclei are shown in Figs. 9 and 10.

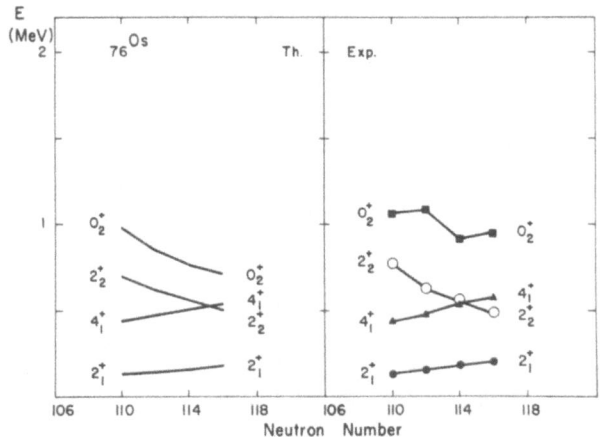

Fig. 9. Typical features of the transitional class B[7]. Energies.

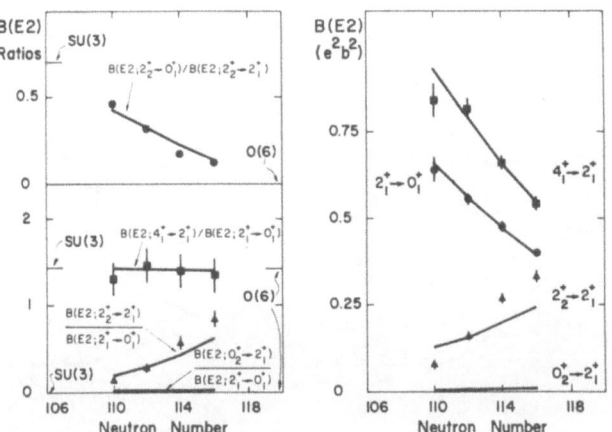

Fig. 10. Typical features of the transitional class B[7]. Electro-
          magnetic transition rates.

Note the rise of the state 2_2^+, the consistently high 0_2^+ state, and the small value $B(E2; 0_2^+ \rightarrow 2_1^+)$. The transitional class C has not been analyzed yet.

In performing a phenomenological analysis of the spectra, such as those in Refs. 6 and 7, we have found it convenient to rewrite the Hamiltonian (2.1) as

$$H = \epsilon n_d + a_o P \cdot P + a_1 L \cdot L + a_2 Q \cdot Q + a_3 T_3 \cdot T_3 + a_4 T_4 \cdot T_4 \quad . \qquad (2.5)$$

The operators P, L and Q are the pairing, angular momentum and quadrupole operators[6]. The operators T_3 and T_4 are boson octupole and hexadecapole operators. In the parametrization (2.5) the three limiting cases of above correspond to

I) SU(5) : $a_o = 0$, $a_2 = 0$, $\qquad\qquad\qquad\qquad\qquad\qquad$ (2.6)

II) SU(3) : $\epsilon = 0$, $a_o = 0$, $a_3 = 0$, $a_4 = 0$, $\qquad\qquad\qquad$ (2.7)

III) O(6) : $\epsilon = 0$, $a_2 = 0$, $a_4 = 0$. $\qquad\qquad\qquad\qquad$ (2.8)

The studies performed so far indicate that while electromagnetic transition rates are usually described rather well, discrepancies occur between calculated and experimental energy levels in nuclei around the O(6) limit. The order of magnitude of these discrepancies is visible in Figs. 6 and 9.

Finally, since collective states in heavy even-even nuclei have usually been described in terms of the geometrical model of Bohr and Mottelson[8], one would like to study the correspondence between the two descriptions. Not much work has been done in this direction. Loosely speaking, one can say that the three limiting cases of the interacting boson model correspond to: I) the anharmonic vibrator II) the axial rotor with degenerate β and γ bands and III) the displaced γ-unstable vibrator. An interesting development in this respect has been provided by Castaños, Chacon, Frank and Moshinsky who have rewritten the interacting boson Hamiltonian (2.1) directly in terms of shape variables $\alpha_\mu, \bar{\alpha}$ and their conjugate momenta $\pi, \bar{\pi}$. α_μ and π_μ are the usual Bohr variables, while $\bar{\alpha}, \bar{\pi}$ are the variables associated with the s-boson. By comparing the first quantized version of (2.1) with the Hamiltonian of the geometrical description, one can see several differences. These are due to (i) the presence of the s-boson, (ii) the finite dimensionality of the boson space and (iii) the conservation of boson number. It is not clear at present how significant these differences are and this problem must be further investigated.

3. THE APPROXIMATION

The model of the previous section has been introduced phenomenologically. Because it appears to provide a classification scheme
in which the various situations encountered in nuclei can be accommo-
dated, one may wonder what is its relation to the microscopic shell-
model. It was pointed out by Talmi that the properties of the s-
and d-bosons are very similar to those of nucleon pairs. It was
then suggested[10] that there is a correspondence between correlated
nucleon pairs and s and d bosons. The s-bosons correspond to nucleon
pairs coupled to L=0, while the d-bosons correspond to pairs coupled
to L=2. Introducing fermion creation operators ($a_{jm}^\dagger$) one can write
the correlated pair operators as

$$s^\dagger = \sum_j \alpha_j s_j^\dagger \qquad\qquad s_j^\dagger = (a_j^\dagger \times a_j^\dagger)^{(0)}$$

$$D_\mu^\dagger = \sum_{jj'} \beta_{jj'} D_{jj';\mu}^\dagger \qquad\qquad D_{jj';\mu}^\dagger = (a_j^\dagger a_{j'}^\dagger)_\mu^{(2)} \qquad (3.1)$$

With this correspondence in mind, we speculated[11] that one can <u>derive</u>
explicitly all coefficients appearing in the boson Hamiltonian, transition operators, etc., by performing the following two steps: (i)
truncate the shell model space from the complete (very large) valence
space (dimensions of the order 10^{12}) to the subspace spanned by the S
and D pairs; (ii) map this space into a boson space as shown in Fig.11.

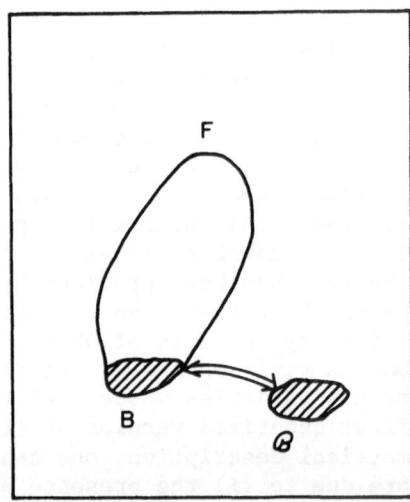

Fig. 11. A schematic representation of the procedure suggested in
         order to construct the boson Hamiltonian and transition op-
         erators. F is the fermion valence space, B is the fermion
         S-D pair space, and $\mathcal{B}$ is the boson s-d space.

In carrying out this program one needs to (i) give a procedure how to calculate the α_j and $\beta_{jj'}$ and (ii) once these are given, how to perform the mapping from the fermion space to the boson space. The only case discussed so far[11] is that of a single large valence shell, $j = {}^{31}/2$, which will be described by Otsuka in his talk. Although this case is certainly not completely realistic, it is interesting to note that the results are already in qualitative agreement with experiment. This is shown in Fig. 12 where energy levels of even-even nuclei in the shell 50-82 are calculated simultaneously, for fixed proton number, $n_\pi = 6$, and varying neutron number, $0 \leqslant n_\nu \leqslant 32$.

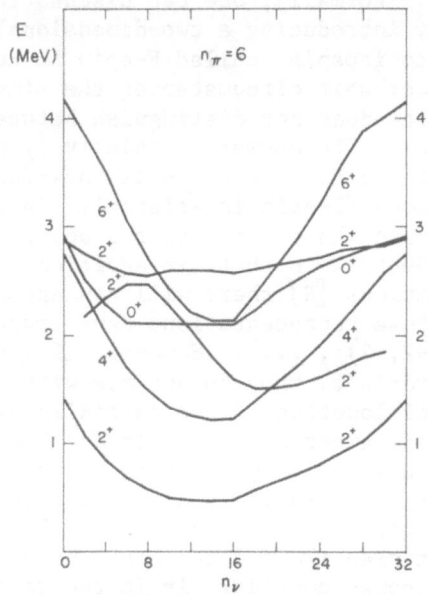

Fig. 12.   Energy spectra of even-even nuclei, for fixed proton number, $n_\pi = 2N_\pi = 6$, and varying neutron number, $0 \leqslant n_\nu \leqslant 32$, in the single j-orbit approximation[11].

The procedure discussed in Ref. 11 allows, given a nucleon-nucleon interaction and the degeneracy of the shell to calculate the spectrum. The results of Fig. 12 have been obtained using a δ-function interaction between identical nucleons and a quadrupole-quadrupole interaction between protons and neutrons.

It is surprising to observe that situations similar to those discussed in the previous section arise from approximate solutions of the shell model within a major shell. In Fig. 12 one can observe a region, $n_\nu \simeq 4$, with a spectrum similar to that of the limiting case I, SU(5), a region, $n_\nu \approx$ 10-14, where the spectrum tends to be like that of II, SU(3), and finally a region, $n_\nu \approx 28$, where the spectrum resembles that of III, O(6). The complicated interweaving of states which is observed experimentally thus appears to be a consequence of the gross shell structure and not of the details of the single particle levels which form the major shell.

From the phenomenological point of view, the most important consequence of having identified the s and d bosons with nucleon pairs is that we must recognize that we have two types of bosons, corresponding to neutron and proton pairs respectively. The total boson Hamiltonian must then be written as $H=H_\pi+H_\nu+V_{\pi\nu}$, where $H_{\pi(\nu)}$ have the same structure as in (2.1), while $V_{\pi\nu}$ contains the proton-neutron interaction. Formally, one can discuss the additional degree of freedom by introducing a two-dimensional variable, similar, but not identical, to isospin, called F-spin in Refs. 10, 11. One may then inquire under what circumstances the simpler version of the model in which one does not distinguish between proton and neutron bosons arises. The answer is relatively simple. This will happen whenever the Hamiltonian H is invariant under proton-neutron transformations (F-spin invarience). In this case, the only difference between the proton-neutron boson model and the simpler version of Sect. 2 is that, in addition to the totally symmetric representations [N] there will now appear other SU(6) representations. These representations have lower symmetry character [N-1, 1] , [N-2, 2] , ... However, if the combined Hamiltonian is F-spin invariant, they do not mix with the representation [N] . The study and location of the partially symmetric states, both theoretically and experimentally, is another problem which has not been analyzed yet. From the single j calculations reported in Fig. 12, it seems that F-spin is not broken whenever both protons and neutrons are particle-like (or hole-like), while it is broken whenever one is particle-like and the other hole-like. However, this situation may change considerably in the realistic case of many non-degenerate j shells.

The combined proton-neutron Hamiltonian $H=H_\pi+H_\nu+V_{\pi\nu}$ provides a tool for more detailed analysis of the spectra. However, from the phenomenological point of view, the full Hamiltonian H, including all parameters which appear in H_π, H_ν and $V_{\pi\nu}$, may be too difficult to use. We have therefore suggested[11] to make use of a simpler version given by

$$H = \varepsilon(n_{d\pi} + n_{d\nu}) + \kappa \, Q_\pi^{(2)} \cdot Q_\nu^{(2)} + a \, M \qquad (3.2)$$

$$Q_{\pi(\nu)}^{(2)} = (d^\dagger \times s + s^\dagger \times \tilde{d})_{\pi(\nu)}^{(2)} + \chi_{\pi(\nu)} \, (d^\dagger \times \tilde{d})_{\pi(\nu)}^{(2)} \qquad (3.3)$$

where $n_{d\pi}$ ($n_{d\nu}$) are the proton (neutron) boson number operators
and M is the Majorana operator. The Majorana operator separates
the symmetric states from the others and it has very small effect
on the low-lying spectrum. The phenomenology of the low-lying
states is thus described, in the parametrization (3.2), (3.3),
by four parameters ϵ, κ, χ_π, χ_ν. Moreover, these parameters are
directly related to the underlying shell model structure. In the
single-j shell approximation of Ref. 11, for example, one can
derive the expected dependence of ϵ, κ, χ_π, χ_ν on the neutron
(N_ν) and proton (N_π) pair numbers

$$\epsilon = \epsilon^{(0)}$$

$$\kappa = \kappa_\pi \kappa_\nu \quad ; \quad \kappa_\pi = \sqrt{\frac{\Omega_\pi - N_\pi}{\Omega_\pi - 1}}\ \kappa_\pi^{(0)} \quad ; \quad \kappa_\nu = \sqrt{\frac{\Omega_\nu - N_\nu}{\Omega_\nu - 1}}\ \kappa_\nu^{(0)} \ ,$$

$$\kappa_\pi \chi_\pi = \left(\frac{\Omega_\pi - 2N_\pi}{\Omega_\pi - 2}\right) \kappa_\pi^{(0)} \chi_\pi^{(0)} \quad ; \quad \kappa_\nu \chi_\nu = \left(\frac{\Omega_\nu - 2N_\nu}{\Omega_\nu - 2}\right) \kappa_\nu^{(0)} \chi_\nu^{(0)} \ . \tag{3.4}$$

Here $\epsilon^{(0)}$, $\kappa_{\pi(\nu)}^{(0)}$, $\chi_{\pi(\nu)}^{(0)}$ are some appropriate constants and $\Omega =$
j + ½ is the pair degeneracy of the shell. One interesting pro-
perty of the expressions (3.4) is that while the $\kappa_{\pi(\nu)}$'s are always
of the same sign, $\chi_{\pi(\nu)}$ change sign in the middle of the shell,
N = $\Omega/2$. In the more general case of many non-degenerate j shells,
the dependence on neutron and proton numbers will certainly be more
complicated, reflecting the underlying detailed shell structure.
A phenomenological analysis of the spectra, using (3.2) and (3.3)
will provide, in addition to a description of their properties,
detailed information on the microscopic structure of the proton
and neutron bosons.

This phenomenological analysis has already been started,
using a program, called NPBOS, written by Otsuka (also available
on request). An example of results is shown in Fig. 13. Others
will be shown by Scholten in his talk. The parameters ϵ, κ,
χ_π, χ_ν determined so far from experiment appear to have the following properties: ϵ and κ vary very smoothly with pair numbers
N_π, N_ν and in fact they are approximately constant in a major
shell; on the contrary χ_π and χ_ν show large variations, with
pronounced subshell effects at 40, 64, 100 and 114. A sketch
of the situation is shown in Fig. 14.

It should be remarked here that once ϵ, κ, χ_π, χ_ν have been
determined as a function of the proton and neutron numbers, that
is once the curves in Fig. 14 have been completely determined by
analyzing the spectra of a set of isotopes and of isotones, the
spectrum of any other even-even nucleus can be calculated.

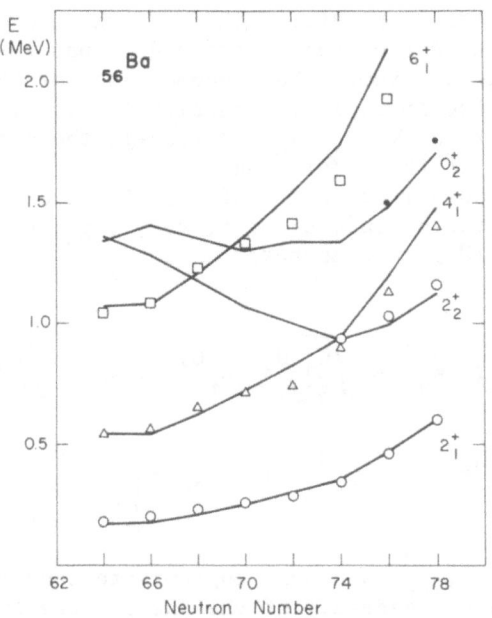

Fig. 13. Calculated energy spectra in the barium isotopes[11].

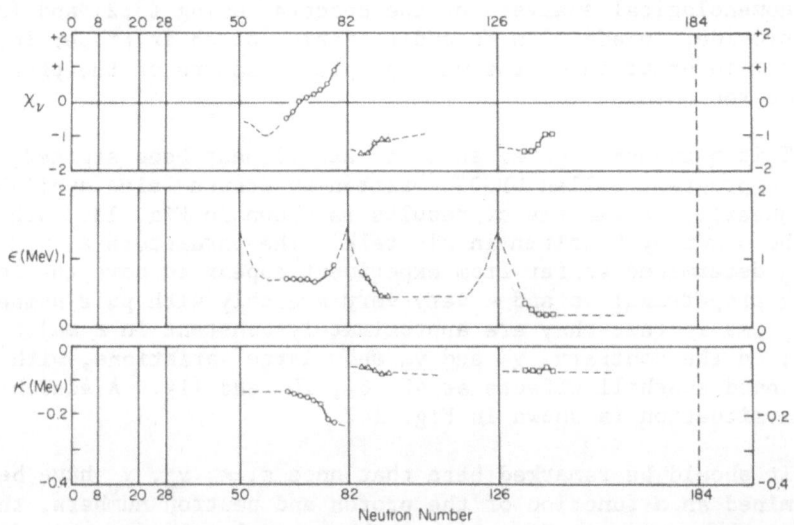

Fig. 14. Approximate behavior of ϵ, κ and χ_ν as a function of
         neutron number. The broken lines indicate extrapolated
         values.

4. CONCLUSIONS.

In conclusion, I have summarized the present status both of the interacting boson model and of the interacting boson approximation. From an experimentalist point of view, the interacting boson model can be used at three different levels of sophistication:

(i) the first level is that in which one compares the energy levels and electromagnetic transition rates with the analytic expressions which correspond to the limiting cases I, II and III. The equivalent of this level in the usual geometrical description is that of looking whether or not the energy levels go like $J(J+1)$. One important new feature here is that all bands, including the excited ones, are given in terms of the same analytic expression while in the geometrical description each band is treated separately.

(ii) the second level is that in which the spectra are analyzed using the program PHINT, which does not distinguish between proton and neutron bosons. The geometrical equivalent of this is to do calculations using a potential energy surface which is an arbitrary function of β and γ, $V(\beta,\gamma)$, and diagonalize the Bohr Hamiltonian as done by Kumar and Baranger.

(iii) finally, at the third level, the spectra are analyzed using the program NPBOS, which takes into account possible differences between protons and neutrons. There is no geometrical equivalent of this level.

From a theorist point of view, the most outstanding problem is that of devicing a simple and yet detailed procedure to calculate the expansion coefficients α_j, β_{jj}, appearing in Eq. (3.1), in the realistic case of many non-degenerate j-shell. This is perhaps the last missing step in providing a link between the shell model and collective descriptions of nuclei.

REFERENCES

1. A. Arima and F. Iachello, Phys. Rev. Lett. <u>35</u> (1975), 1069.
2. A. Arima and F. Iachello, Ann. Phys. (N.Y.) <u>99</u> (1976), 253.
3. A. Arima and F. Iachello, Ann. Phys. (N.Y.) <u>111</u> (1978), 201.
4. A. Arima and F. Iachello, Phys. Rev. Lett. <u>40</u> (1978), 385.
5. J. A. Cizewski, R. F. Casten, G. J. Smith, M. L. Stelts, W. R. Kane, H. G. Börner and W. F. Davidson, Phys. Rev. Lett. <u>40</u> (1978), 167.
6. O. Scholten, F. Iachello and A. Arima, Ann. Phys. (N.Y.) <u>115</u> (1978), 321.
7. R. F. Casten and J. A. Cizewski, Phys. Lett. <u>79B</u> (1978), 5.
8. A. Bohr and B. R. Mottelson, "Nuclear Structure", Vol. II, W. A. Benjamin, Reading, Mass. (1975).

9. O. Castanõs, E. Chacón, A. Frank and M. Moshinsky, to be pub-
    lished in J. Math. Phys.

10. A. Arima, T. Otsuka, F. Iachello and I. Talmi, Phys. Lett. 66B
    (1977), 205.

11. T. Otsuka, A. Arima, F. Iachello and I. Talmi, Phys. Lett. 76B
    (1978), 139; T. Otsuka, A. Arima and F. Iachello, Nucl. Phys.
    A309 (1978), 1.

A PHENOMENOLOGICAL STUDY OF EVEN-EVEN NUCLEI

IN THE NEUTRON-PROTON I.B.A.

O. Scholten

Kernfysisch Vernsneller Instituut

University of Groningen, Groningen, the Netherlands

1. INTRODUCTION

In this talk, I will give an overview of the phenomenologic calculations that have been performed in the framework of the I.B.A. model. All the calculations that I will present were done in the neutron-proton I.B.A. model. The main interest in these calculations was to extract from the experimental data the parameters appearing in the Hamiltonian, which could then later be compared with those arising from the microscopic calculations[1]. For this reason, we did not try to get the best possible fit for each individual nucleus but rather attempted to reproduce the observed trends in excitation energies, binding energies, B(E2) etc. for a chain of isotopes or isotones.

The outline of the I.B.A. model has already been given in the preceeding talk. I will repeat only shortly those points which are relevant to the present talk. The Hamiltonian used in the calculations to be reported below is,

$$H = \epsilon \ (n_{d_\nu} + n_{d_\pi}) + \kappa \ \ Q_\nu \cdot Q_\pi + aM \qquad (1.1)$$

where

$$Q_{\nu(\pi)} = (s^\dagger \times \tilde{d} + d^\dagger \times s)^{(2)}_{\nu(\pi)} + \chi_{\nu(\pi)} \ (d^\dagger \times \tilde{d})^{(2)}_{\nu(\pi)} \qquad (1.2)$$

and

$$M = (s^\dagger_\nu \times d^\dagger_\pi - d^\dagger_\nu \times s^\dagger_\pi)^{(2)} \cdot (s_\nu \times \tilde{d}_\pi - \tilde{d}_\nu \times s_\pi)^{(2)} +$$

17

$$+ \xi \sum_{k=1,3} (d_\nu^\dagger \times d_\pi^\dagger)^{(k)} \cdot (\tilde{d}_\nu \times \tilde{d}_\pi)^{(k)} \tag{1.3}$$

The important pieces here are: the boson energy ε, and the strength of the neutron-proton quadrupole-quadrupole force, κ, together with χ_ν and χ_π. The Majorana force M merely serves to push up the states that are antisymmetric in the neutron and proton degrees of freedom, since there is experimentally no evidence for states of this kind in the low-lying part of the spectrum. To this H one could add a neutron-neutron and proton-proton interaction, if one would wish to obtain a more detailed description of the experimental data. Using (1.1), it is possible to obtain spectra which are similar to those of the I.B.A. model with only one kind of boson. The SU(5)[2] or anharmonic vibrator limit is obtained when $\varepsilon \gg \kappa$, the SU(3)[3] or axial rotor limit when $\varepsilon \ll \kappa$ and $\chi_\nu = \chi_\pi = -\frac{1}{2}\sqrt{7}$, and the 0(6)[4] or γ-unstable limit when $\varepsilon \ll \kappa$ and $\chi_\nu = -\chi_\pi$. In the actual calculations, when considering a chain of isotopes, the parameters ε, κ and χ_ν were allowed to vary in a smooth way from nucleus to nucleus, while χ_π was kept fixed.

In the neutron-proton I.B.A. the T(E2) operator has the form

$$T(E2) = e_\nu \left[(s^\dagger \times \tilde{d} + d^\dagger \times s)_\nu^{(2)} + \chi_\nu (d^\dagger \times \tilde{d})_\nu^{(2)} \right] +$$

$$+ e_\pi \left[(s^\dagger \times \tilde{d} + d^\dagger \times s)_\pi^{(2)} + \chi_\pi (d^\dagger \times \tilde{d})_\pi^{(2)} \right] \tag{1.4}$$

In our calculations, the effective boson charges e_ν, e_π were taken equal, $e_\nu = e_\pi = e$. The constants χ_ν and χ_π could in principle be different from the χ_ν and χ_π appearing in H. However, they were taken equal in order to reduce the number of free parameters. The boson effective charge e was chosen as to give the correct normalization for the B(E2) values. For a given chain of isotopes, it was kept fixed. Branching ratio's are independent of this parameter.

As it was mentioned in the preceeding talk, it is easy to make predictions on how the parameters in the Hamiltonian should vary within a major shell, when the single particle levels in this major shell can be considered to be degenerate and the shell, therefore, can be approximated with a single j-shell. For the more general case, I refer to the talk given by Otsuka. In the degenerate case, one expects that ε and κ remain essentially constant, but that χ will change sign, when crossing the middle of the shell. In reality the situation will be more complicated, but one expects that ε and κ will be approximately constant and the only parameter that will change much is χ. This agrees with the phenomenological results.

The isotopes for which we have done calculations are:

                    1st the Ba-isotopes,
                    2nd the Sm-isotopes,
                    3rd the Os-isotopes,
                    4th the Th-isotopes, and as last
                    5th the N=90 isotones

I will discuss each of these chains of isotopes and isotones in
more detail in the following sections.

2.   THE BARIUM-ISOTOPES

        These calculations have been done by Otsuka and Puddu. The
Ba isotopes show quite nicely a transition from an almost SU(3)
(axial rotor) spectrum for N≈66 to an 0(6) (γ-unstable) spectrum
for N≈76. For N=66, the energies of the levels in the ground-state
band go approximately like L(L+1) and there is no observed low-
lying 0_2^+ or 2_3^+ state. For N=76, however, the ratio of excitation
energies for the 4_1^+ and 2_1^+ level is well below the rotational value
of 3.3, and what is more important, there is a 2_2^+ level at approximately the energy of the 4_1^+ state, while the 0_2^+ level is lying
about 500 keV higher. The B(E2) values give some more indications
of this change, but I will discuss them later. The result of the
calculations is shown in Fig. 2.1 and the fitted parameters in

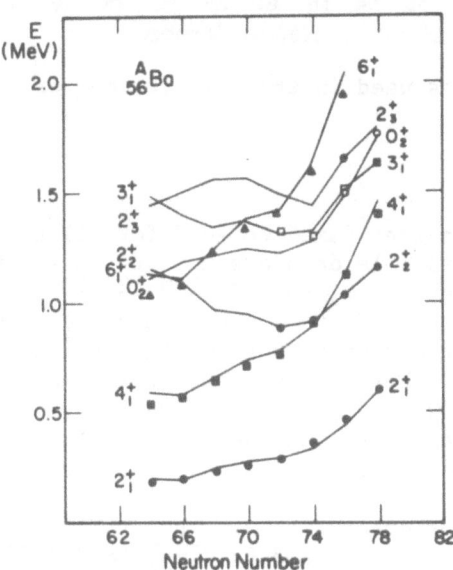

Fig. 2.1   Calculated (lines) and experimental[7],[8] (points)
            excitation energies for the Ba-isotopes.

Fig. 2.2. It is important to notice that ϵ and κ stay all the
time approximately constant, while χ_ν is changing sign as one would
expect on the basis of the "single-j-shell" predictions.

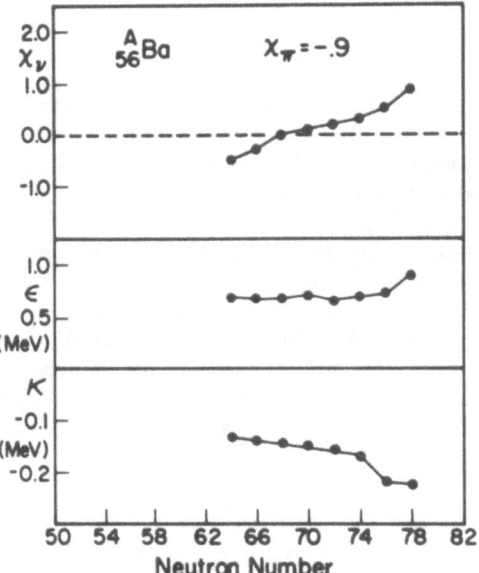

Fig. 2.2 Parameters used in the calculation for the Ba-isotopes.

The value of χ_π was kept fixed to -0.9 for all the isotopes. In
some cases, the available data are not sufficient to determine the
three free parameters ϵ, κ, χ_ν. In ^{120,122,124,126}Ba, for example,
only levels of the ground state band are known. For these cases,
the parameters were obtained by a smooth continuation of the parameters found for the heavier isotopes (for ϵ and κ this is just a
straight line) or by taking, as was done for χ_ν, the value from
another isotope chain. For the heavier isotopes, the parameters
could be fixed much better.

Other important quantities are the B(E2) values. The calcu-
lated and measured values for B(E2;$2_1^+\rightarrow0_1^+$) and B(E2;$4_1^+\rightarrow2_1^+$) are
plotted in Fig. 2-3. The only free parameter in the calculation,
the boson effective charge, was fixed to the B(E2;$2_1^+\rightarrow0_1^+$) in ^{130}Ba,
and kept constant for the other Ba isotopes, $\epsilon = 0.141$ eb. The
decrease in the B(E2) when going from ^{126}Ba to ^{134}Ba is reproduced

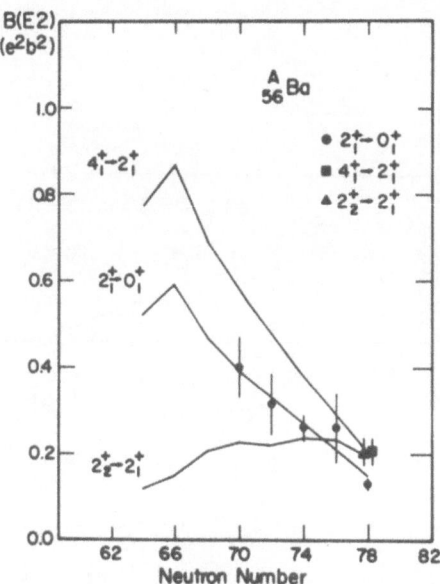

Fig. 2.3 Calculated (lines) and experimental[8,9] (points)
          B(E2) values for the $2_1^+\to0_1^+$, $4_1^+\to2_1^+$, and $2_2^+\to2_1^+$ transitions
          in the Ba isotopes.

very well. This decrease is mainly due to the decrease of the
number of bosons. An equally important quantity is the ratio R =
$B(E2;4_1^+\to2_1^+)/B(E2;2_1^+\to0_1^+)$. Since there exists for this ratio only
one data point, I will not show the corresponding figure. The
theory predicts a constant value for R because in the SU(3) as
well as in the O(6) limit R should be equal to 10/7. Fig. 2.4
shows some branching ratios. The ratio $R_1= B(E2;2_2^+\to0_1^+)/B(E2;2_2^+\to2_1^+)$
should vanish in the O(6) limit because the numerator vanishes
($\Delta\tau=2$ transitions are not allowed) while the denominator becomes
equal to $B(E2;4_1^+\to2_1^+)$. In the SU(3) limit on the other hand, both
the numerator and denumerator become small, but their ratio will
stay finite $R_1(SU(3)) = 0.7$. The same situation holds for $R_2 =$
$B(E2;3_1^+\to2_1^+)/B(E2;3_1^+\to4_1^+)$. In the SU(3) limit both transitions are
forbidden, but the ratio is finite; $R_2(SU(3)) = 2.5$ while in the
O(6) limit, the numerator vanishes ($\Delta\tau=2$) and the denominator is
comparable with $B(E2;4_1^+\to2_1^+)$.

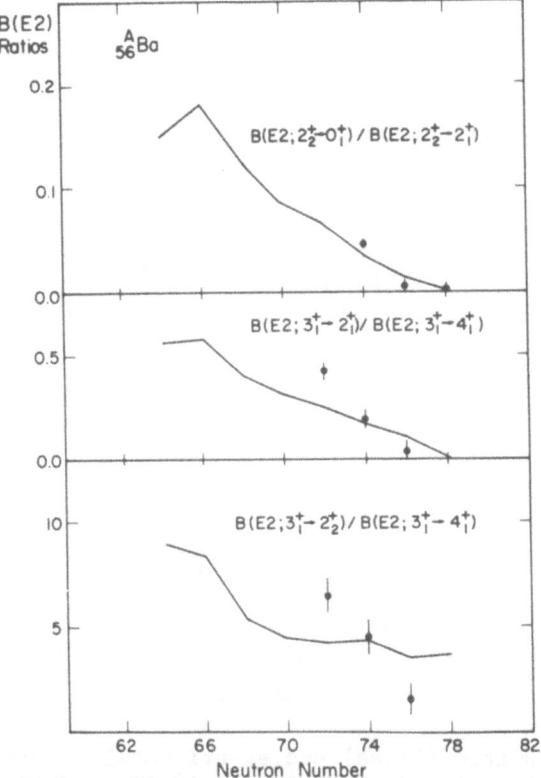

Fig. 2.4 Some calculated (lines) and experimental[8] (points)
         B(E2) branching ratios in the Ba-isotopes.

        The results for the quadrupole moments of the 2_1^+ state are shown
in Fig. 2.5. In the calculation of the quadrupole moment, there is
no free parameter to be adjusted, since the boson effective charge
had already been fixed by equating the calculated value to the observed $B(E2;2_1^+ \to 0_1^+)$ value in ^{130}Ba. Unfortunately the measured
values scatter considerably. Some more precise data would be wel-
come.

        Other calculations show that similar fits as presented here can
be obtained for the neighbouring Ce and Xe isotopes, using the same
parameters as for Ba. For the Xe isotopes, however, the effect of
the interaction between equal particles, neglected in the calculations for Barium presented here will be more important. The reason
is that in the case of Xe there are only two proton-bosons, which
will give rise to less collectivity or, in other words, less dominance of the neutron-proton interaction than in the case of the Ba-
isotopes.

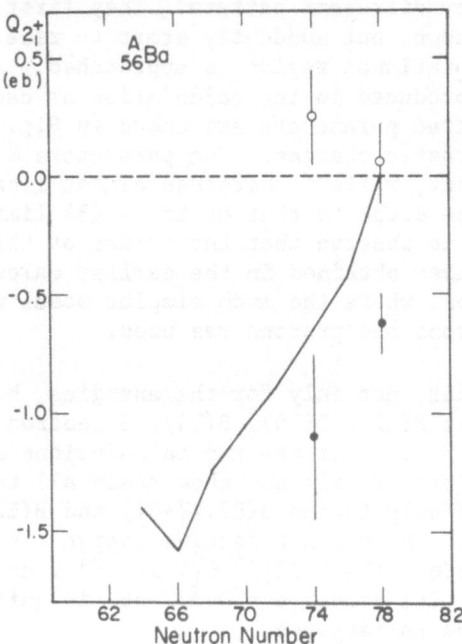

Fig. 2.5   Calculated (line) and experimental[8] (points) quadrupole
moments for the 2_1^+ state in the Ba-isotopes. The large
discrepancy between the different measured values might
be due to a different sign for the P_3 interference term.

3.   THE SAMARIUM ISOTOPES

These calculations have been done by Otsuka, deJong and myself.
The Sm isotopes are the "classical" example of a transition between
a vibrational spectrum (^{148}Sm) and the spectrum of a good axial rotor
(^{156}Sm). Many calculations, using other methods, have been done for
these nuclei. Some are given in ref. 5.

The transition from SU(5) (vibrational) to SU(3) (rotational)
manifests itself in several ways. In the vibrational limit, the
0_2^+, 4_1^+ and 2_2^+ level lie quite close together at approximately two
times the energy of the 2_1^+ state, while in the rotational limit,
the energies of the states in the ground-state band behave like
L(L+1) with the 0_2^+ and 2_2^+ relatively high in the spectrum (for
example in ^{154}Sm, they lie at more than ten times the energy of the
2_1^+ state). In the transition from SU(5) to SU(3), the 2_1^+ level
drops smoothly. The other levels from the ground state band also
follow this behaviour. On the contrary, the 0_2^+ state and the 2_2^+

state follow a very different pattern. They first follow the 4^+_1
level on the way down, but suddenly start to rise again (near
^{152}Sm) when the rotational region is approached. All these features
are quite well reproduced in the calculation as can be seen in
Fig. 3.1. The fitted parameters are shown in Fig. 3.2. None of the
parameters show drastic changes. The parameters κ and χ_ν remain
essentially constant, while ε decreases almost linearly. Both χ_ν
and χ_π have a value close to that of the SU(3) limit ($-\sqrt{7}/2\approx-1.3$).
It is interesting to observe that the values of the parameters are
very similar to those obtained in the earlier calculation[6] of this
transitional region, where the much simpler model with no distinction between neutrons and protons was used.

It appears that, not only for the energies, but also for the
other quantities as B(E2), B(E0), B(E4), 2 neutron separation
energies etc. the results of the two calculations are practically
identical. Therefore, I will not show again all these results,
but confine myself only to the $B(E2;2^+_1\rightarrow0^+_1)$ and $B(E2;4^+_1\rightarrow2^+_1)$ values
shown in Fig. 3.3. The boson effective charge $e = e_\pi = \bar{e}_\nu$ was
chosen as to reproduce the $B(E2;2^+_1\rightarrow0^+_1)$ in ^{154}Sm and is kept constant throughout. Its value, $e = 0.132$ eb. is quite close to the
value found for the Ba-isotopes.

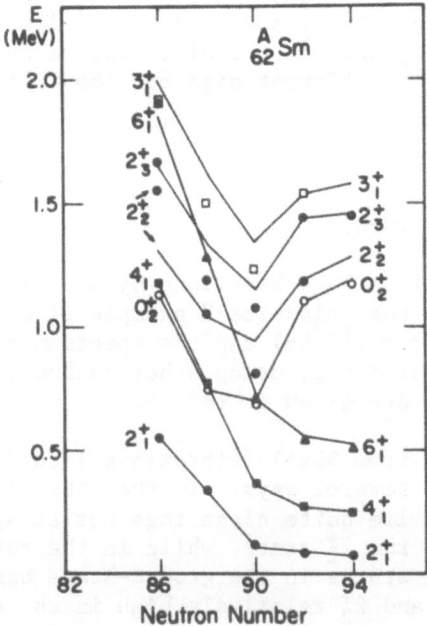

Fig. 3.1 Calculated (lines) and experimental[7] (points) excitation
energies for the Sm-isotopes.

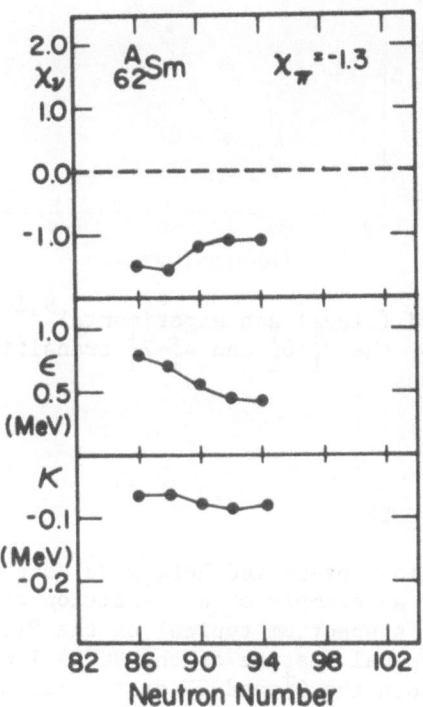

Fig. 3.2 Parameters used in the calculation for the Sm-isotopes.

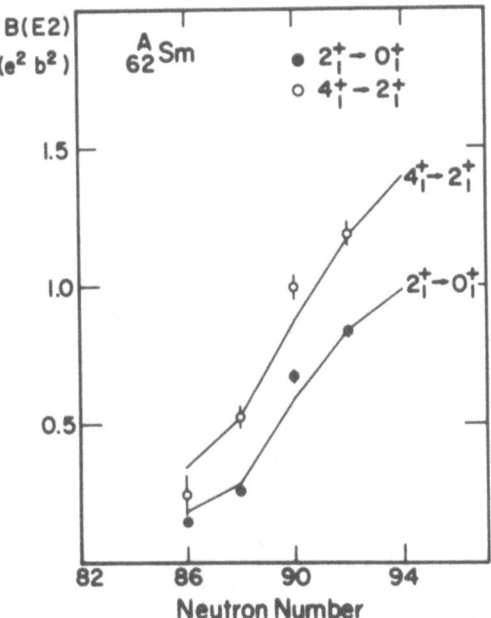

Fig. 3.3 Calculated (lines) and experimental[9,10] (points) B(E2)
values for the $2_1^+ \rightarrow 0_1^+$ and $4_1^+ \rightarrow 2_1^+$ transition in the
Sm-isotopes.

4. THE OSMIUM ISOTOPES

These calculations presented here were done by Spanhof. The
Os isotopes provide an example of a transition from a SU(3)-like
spectrum (^{184}Os) to a spectrum typical of the O(6) limit (^{192}Os),
see Fig. 4.1. A typical feature of the SU(3) limit is, as it was
said before, that both the 0_2^+ and 2_2^+ states lie at a relatively
high excitation energy, while in the O(6) limit the 2_2^+ state is
much lower than the 0_2^+ state, approximately at the position of the
4_1^+ level. This change in structure is very well reproduced by the
calculation. The parameters obtained are shown in Fig. 4.2. Again,
ε and κ are practically constant for all the isotopes. The only
parameter which really changes is χ_ν. In the exact SU(3) limit,
$\chi = -\frac{1}{2}\sqrt{7} = -1.3$. The fact that it is only -0.4 here means that for
^{184}Os the SU(3) symmetry is not yet fully developed (the 2_2^+ state is
still below the 0_2^+). In order to have spectra with an O(6) symmetry
the sign of χ_ν and χ_π must be opposite. Thus χ_ν becomes positive
for ^{192}Os. This behaviour of χ_ν is consistent with the predictions
of the single degenerate j-shell: at the end of a major shell χ
should become positive. On the other side, the negative value of
χ_π which is needed to fit the data, is inconsistent with the assump-

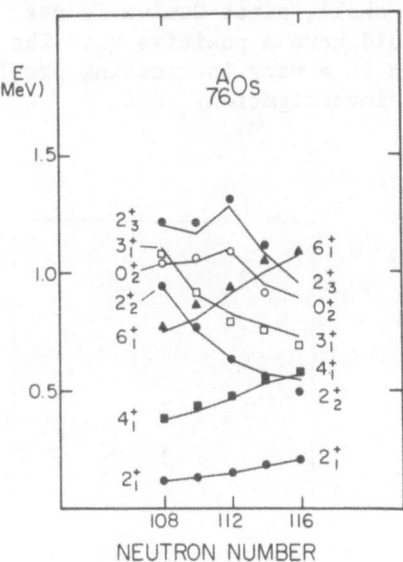

Fig. 4.1 Calculated (lines) and experimental[7] (points) excitation
energies for the Os-isotopes.

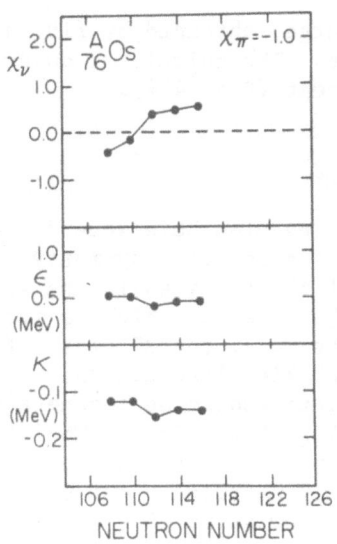

Fig. 4.2 Parameters used in the calculation for the Os-isotopes.

tion that the proton levels in the major shell 50-82 behave like a
single degenerate j-shell, since Osmium is near the end of the shell
(Z=76) and thus should have a positive χ_π. The anomalous behaviour
of χ_π in this region is a very interesting problem which requires
further theoretical investigation.

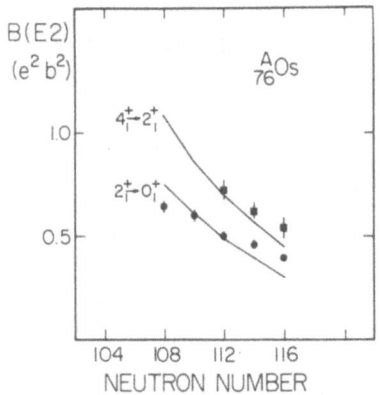

Fig. 4.3 Calculated (line) and experimental[9] (points) values
          for $B(E2;2_1^+\rightarrow0_1^+)$ and $B(E2;4_1^+\rightarrow2_1^+)$ values in the Os-
          isotopes.

The boson effective charge, obtained by fitting the B(E2) value
in ^{186}Os is e = 0.14 eb. The calculated quadrupole moments also
agree well with experiment (Fig. 4.4).

5. THE THORIUM-ISOTOPES

The calculations presented here were done by Alting. The Th
isotopes are examples of good SU(3) symmetry. This is evident from
the L(L+1) behaviour of the energies in the ground-state band and
from the location of the 0_2^+ state which is rather high in the spectrum (Fig. 5.1). Furthermore, the 2^+ state of the "γ band" and the
2^+ state of the "β-band" lie close together. In the exact SU(3)
limit, the 2_2^+ and 2_3^+ state should be degenerate. The parameters
obtained are shown in Fig. 5.2. In the Th-isotopes, both the valence neutrons and protons are in the beginning of a major shell,
giving rise to a strongly negative χ for both neutrons and protons.
This, combined with the low value of ε≈0.2 Mev, gives rise to the
SU(3) like spectra. Comparing the Thorium parameters with those
of Samarium, one finds many similarities: a similar drop in χ_ν,
a rising of ε towards the closed shell, and essentially the same
constant value for κ.

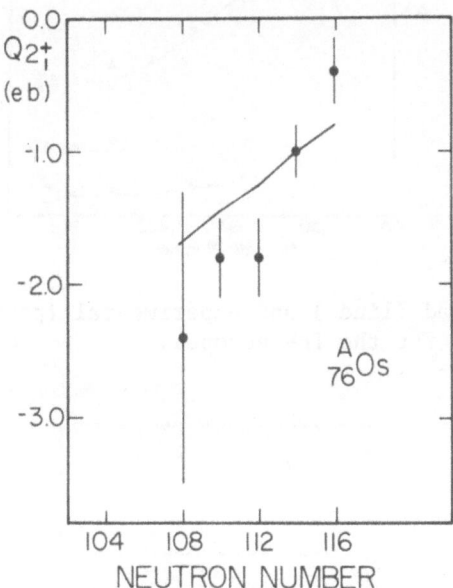

Fig. 4.4 Calculated (line) and experimental[11](points) values for
the quadrupole moment of the 2_1^+ state in the Os-isotopes.

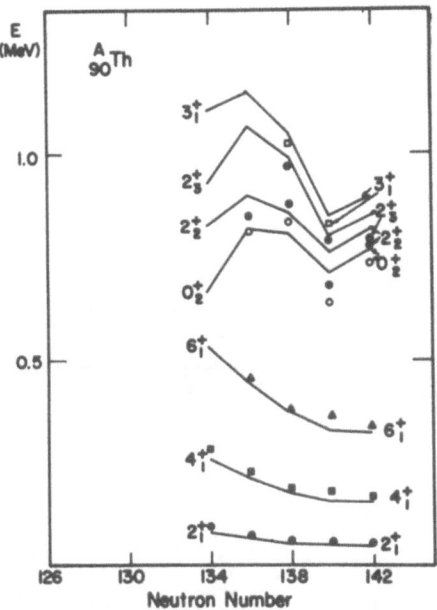

Fig. 5.1 Calculated (lines) and experimental[7] (points) excitation
          energies for the Th-isotopes.

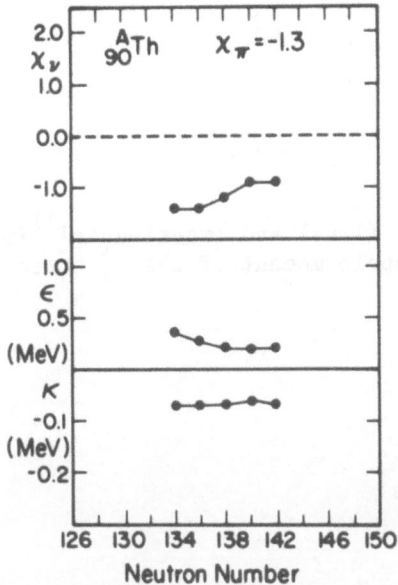

Fig. 5.2 Parameters used in the calculation for the Th-isotopes.

The calculated and measured values for the $B(E2;2_1^+ \to 0_1^+)$ and $B(E2;4_1^+ \to 2_1^+)$ are plotted in Fig. 5.3. The boson effective charge, obtained by adjusting the $B(E2;2_1^+ \to 0_1^+)$ in ^{228}Th is e=0.182 eb., a value much larger than for Sm. The reason for this is the much

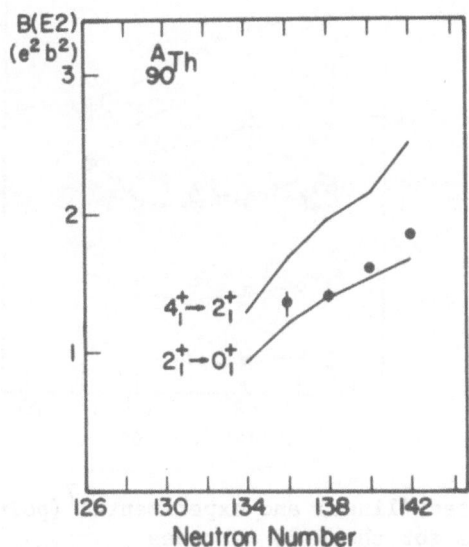

Fig. 5.3.   Calculated (lines) and experimental[9,12] (points) $B(E2)$ values for the $2_1^+ \to 0_1^+$ and $4_1^+ \to 2_1^+$ transition in the Th-isotopes.

larger $\langle r^2 \rangle$ of the valence shells as compared with the Sm-isotopes. Since the structure of the Th-isotopes does not change, the rise in the $B(E2)$ value is much less than for the Sm isotopes. If the isotopes were all the time exactly SU(3), the rise in the $B(E2)$ value would go quadratically with the total number of bosons. When going from ^{224}Th to ^{232}Th this would give an increase by a factor 2.25 which is comparable with experiment.

6.   THE N=90 ISOTONES

These calculations were done by myself. Up to now, we have only studied the behaviour of the parameters as a function of the number of neutrons. It is also interesting to know the change of the parameters as a function of the number of protons. For this study, we choose the N=90 isotones: ^{150}Nd, ^{152}Sm, ^{154}Gd, ^{156}Dy, and ^{158}Er. Their spectra are shown in Fig. 6.1. As can be seen

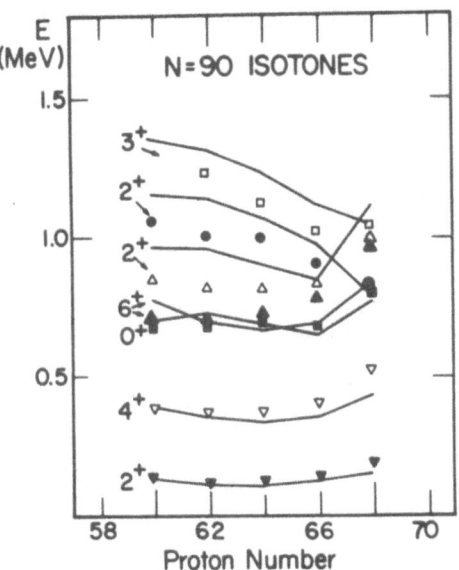

Fig. 6.1 Calculated (lines) and experimental[7] (points) excitation
energies for the N=90 isotones.

from this figure, there is a clear trend of going from a SU(3)-like
spectrum for Nd, Sm, and Gd to that of the 0(6) limit for Er. The
best indication for this change comes from the 2^+ state from the
γ-band. In ^{154}Gd this state is still well above the 2^+_β, for ^{156}Dy
it is at the same energy as the 2^+_β. This rapid change implies a
steep increase in χ_π, as can be seen in Fig. 6.2 The parameters
for ^{152}Sm are of course the same as those given in Fig. 3.2. In
the calculations for the other isotones, we did not try to optimise
ε and just took the value from ^{152}Sm.

The $B(E2;2^+_1 \rightarrow 0^+_1)$ values are plotted in Fig. 6.3. The boson effective charge, kept constant for all the isotones, was e=.140 eb.
The observed maximum in the B(E2) strength is reproduced very well.
The reason for this maximum is that when going from Nd to Gd the
B(E2) increases because the number of bosons increases. When going
from Gd to Dy, the change in structure (from SU(3) to 0(6)) starts
to play a role and there is no further increase. The sharp decrease
for Er has two reasons: i) the change in structure, and ii) the
decrease in the number of bosons (the middle of the major shell is
at Z=66, Dy).

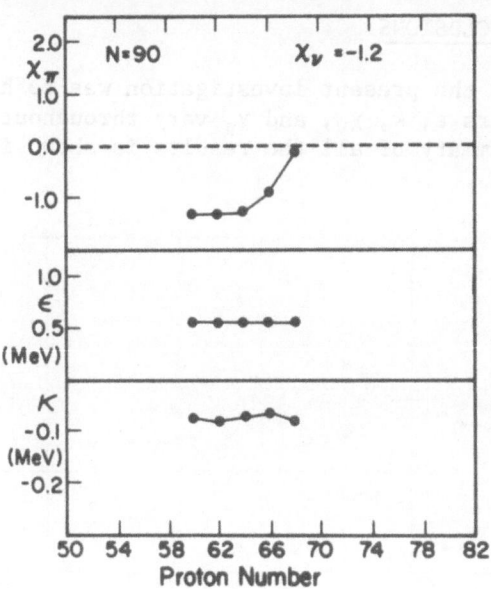

Fig. 6.2   Parameters used in the calculation for the N=90
           isotones.

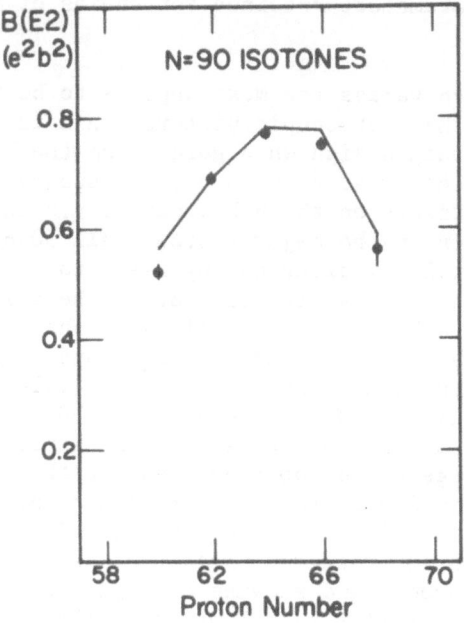

Fig. 6.3   Calculated (line) and experimental[9] (points) values for
           the $B(E2;2_1^+\rightarrow0_1^+)$ in the N=90 isotones.

7. SUMMARY AND CONCLUSIONS

The purpose of the present investigation was to have an idea
of how the parameters ε, κ, χ_ν, and χ_π vary throughout the entire
period table. A summary of all the results is shown in Fig. 7.1.

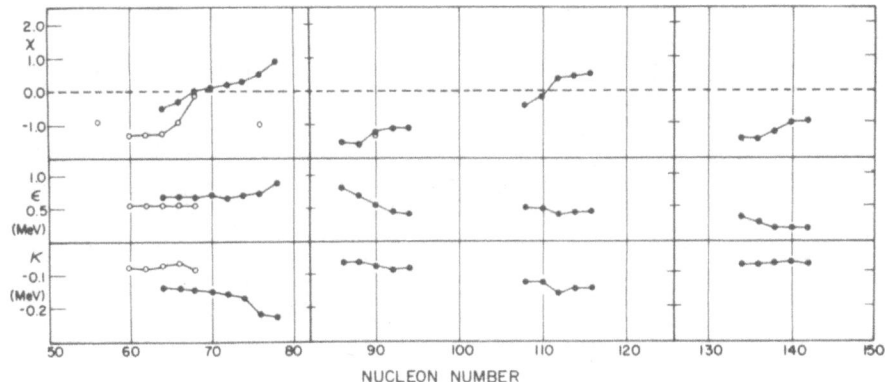

Fig. 7.1   Summary of the values of the parameters χ_ν, χ_π, κ and ε
           as appearing in Eq. 1.1, used in the calculations pre-
           sented in this paper. The filled (open) circles repre-
           sent parameters obtained for chains of isotopes (isotones).

The parameter which varies the most appears to be χ. From the
point of view of the microscopic picture, this is to be expected.
The parameter χ changes sign when going from the beginning to the
end of a shell. The presence of a partial subshell closure can have
therefore large effects on the behaviour of χ. An example of this
may perhaps be seen in the major proton shell 50-82, where χ_π returns
negative (-1.0) at Z=76, after having risen to about zero to Z=70.
Another interesting observation is that the behaviour of the various
parameters seems to be similar in all major shells. For example, ε
is large (1.4 MeV) near the closed shells and drops going away from
the closed shell until it reaches a constant value (about 0.7 MeV
for the 50-82 shell, 0.5 MeV for 82-126, and 0.2 MeV for the major
shell starting at 126). On the basis of a simple microscopic theory
with all levels degenerate, one would expect that ε is constant
(≈1.4 MeV). A more sophisticated microscopic theory shows that this
drop can be accounted for by taking second order contributions from
the neutron-proton interaction. Also the phenomenological values
of κ, being very smooth and showing a slight rise near the closed
shells, are rather similar in all major shells.

To conclude, I would say that the model seems to be able to
describe a large variety of nuclei; from vibrational to rotational,

to γ-unstable, with a single, very simple Hamiltonian. Furthermore, the extracted parameters follow qualitatively the predictions of the most simple microscopic theory.

REFERENCES

1. T. Otsuka, A. Arima, F. Iachello and I. Talmi, Phys. Lett. 76B (1978), 139. T. Otsuka, A. Arima and F. Iachello, Nucl. Phys. A309 (1978), 1.
2. A. Arima and F. Iachello, Ann. Phys. (N.Y.) 99 (1976), 253.
3. A. Arima and F. Iachello, Ann. Phys. (N.Y.) 111 (1978, 201.
4. A. Arima and F. Iachello, Phys. Rev. Lett. 40 (1978), 385.
5. K. Kumar, Nucl. Phys. A231 (1979), 189 and references therein.
6. O. Scholten, F. Iachello and A. Arima, Ann. Phys. (N.Y.) 115 (1978), 321.
7. M. Sakai and A. C. Rester, Atomic Data and Nuclear Data Tables 20 (1977), 441.
8. C. W. Towsley, R. Cook, D. Cline and R.N. Moroshka, Proc. Int. Conf. on Nucl. Moments and Nucl. Structure, Osaka (1972), 442.
    T. T. Simpson, D. Eccleshall, A. T. Yates and N. T. Freeman, Nucl. Phys. A94 (1967), 177.
    T. R. Kerns and T. X. Saladin, Phys. Rev. C6 (1972), 1016.
    W. Walters, private communication.
    D. R. Zolnewski and T. T. Sugihara, Phys. Rev. C16 (1977), 408.
9. S. Raman, W. T. Milner and P. M. Stelson, private communication to F. Iachello.
10. R. M. Diamond, F. S. Stephens, K. Nakai and R. Nordhagen, Phys. Rev. C3 (1971), 344.
    N. Rud, G. T. Ewan, A. Christy, D. Ward, R. L. Graham and T. S. Geiger, Nucl. Phys. A191 (1972), 545.
    R. M. Diamond, G. D. Symons, T. L. Quebert, K. M. Maier, T. R. Leigh and F. S. Stephens, Nucl. Phys. A184 (1972), 481.
11. A. Christy and O. Häusser, Nucl. Data Tables 11 (1972), 281.

THE IBA MODEL IN THE Pt-Os REGION*

R. F. Casten

Brookhaven National Laboratory

Upton, New York 11973

ABSTRACT

It is shown that [196]Pt is an excellent empirical manifestation of the newly proposed, and heretofore unobserved, O(6) symmetry of the Interacting Boson Approximation Model. Further, with this as a starting point, a new, simpler, interpretation is possible of the complex Pt-Os transition region as initiating an O(6) → rotor (SU(3)) transition.

I.   INTRODUCTION

In the past few years Iachello and Arima[1-3] have developed a new model, the Interacting Boson Approximation (IBA) model, which attempts to account for the collective excitations of all heavy even-even nuclei, except those very near to closed shells, in a comprehensive and unified fashion. The model is based on the assumption that the full Fermion Hamiltonian can be successfully truncated to terms involving pairs of particles coupled to spin 0 and 2 and, further, that this truncated Hamiltonian can then be mapped onto an equivalent Boson Hamiltonian containing bosons with angular momentum 0(s) and 2(d). The microscopic structure of these bosons, and the field theoretic status of this approach are treated by others at this conference. I wish to discuss here the phenomenological aspects of the model and their comparison with extensive empirical data in the Os-Pt region.

The IBA model is couched in the language of group theory. Though nuclei with a continuous range of properties can be treated by the model, three particular limiting cases are recognizable through the medium of their group classifications. These three

limits correspond to specific nuclear symmetries. Two of them, the
(anharmonic) vibrator (SU(5)) and the symmetric rotor (SU(3)) have
long been known as are nuclei empirically approximating each. The
third symmetry, the O(6) limit, though evidently contained in com-
plentary geometrical models, was heretofore unrecognized, but occurs
on the same footing in the IBA as the other two.

I would like today to discuss two principal points. The first
is that ^{196}Pt appears to be an excellent, and the first disclosed,
empirical manifestation of the O(6) limit of the IBA. The second is
that this recognition, combined with the structure of the IBA, now
allows a new interpretation of the complex and heretofore unsatis-
factorily understood Pt-Os transition region that avoids the need to
introduce the usual competing and coexisting degrees of freedom such
as oblate, prolate, triaxial, γ-soft and hexadecapole shapes.

First, however, a few brief remarks on the experimental techniques are appropriate, in particular to highlight some of the special
properties of the (n,γ) reaction used. This reaction is chiefly
characterized by an inherent non-selectivity and thus provides a
technique whereby one has access to all states in certain excitation
energy and spin ranges. In particular, the proper use of average
resonance capture techniques specifically exploits the non-selectivity
in such a way that, for example in ^{196}Pt, one observes all 0^+, 1^+ and
2^+ levels below ∿2.5 MeV. More importantly, one is, in fact,
guaranteed of observing all such states and, as a corollary, one
therefore knows that there are no others. This provides an immensely
potent empirical tool as it allows one to test if there is in fact
a one-to-one correspondence between empirical levels and those predicted by a given model. For the O(6) limit, the structure of the
states is such that no other reaction can provide such a test.

Furthermore, with the use of sophisticated devices such as
curved crystal spectrometers, one can detect γ rays over an enormous
dynamic range of intensities and with an energy precision in the few
eV range. This is particularly important in the present case since
a signature of the O(6) limit is sequences of low energy cascade
transitions occurring relatively high in the level scheme. These
represent the largest B(E2) values for the decay of the initial
states but have very weak intensities due to the strong E_γ^5 factor
relating E2 electromagnetic matrix elements to observed intensities.

The (n,γ) experiments were performed both at BNL and at the
Institut Laue Langevin (ILL) in Grenoble, France. They involved
(at BNL) thermal, resonance and average resonance capture experiments
in both singles and coincidence modes using Ge(Li) detectors,
and (at the ILL) the precision measurements of low energy γ rays
(below ∿1.5 MeV) with curved crystal spectrometers following single
and twofold neutron capture and the study (for ^{190}Os) of the (n,e$^-$)
reaction. The data are discussed in detail in refs. 4-8.

II. THE O(6) SYMMETRY

The limits and characteristics of the IBA are treated in detail elsewhere in the literature[1-3] and at this conference. I shall only briefly summarize some of the salient features of the model as it applies to the subsequent discussion. For further details the reader is referred to refs. 1-3 and 9-11.

Figure 1 shows the O(6) symmetry. The eigenvalue equation is

$$E(\sigma, \tau, J) = A(N-\sigma)(N+\sigma+4) + B\tau(\tau+3) + CJ(J+1) \qquad (1)$$

where σ and τ are O(6) quantum numbers and J is the level spin. N is the total boson number ($N = n_s + n_d$) and is given by half the total number of protons and neutrons from their nearest respective closed shells. In this phenomenological form of the IBA no distinction is made between proton and neutron bosons nor between particles and holes. For ^{196}Pt, ^{194}Pt and ^{190}Os, N=6, 7, and 9, respectively. The allowed ranges of σ and τ are summarized in refs. 3, 5, 9 and 10, and are evident for N=6 in Fig. 1. σ is a sort of major quantum number. For each σ value, the same sequence of states appears with the same spins (except for a differing spin cutoff) and spacings but commencing with a different base energy which is given by $A(N-\sigma)x$ $(N+\sigma+4)$. The τ quantum number somewhat resembles a phonon number but the level sequences within a given σ family are not those of the harmonic oscillator. Rather, the allowed levels are those of the

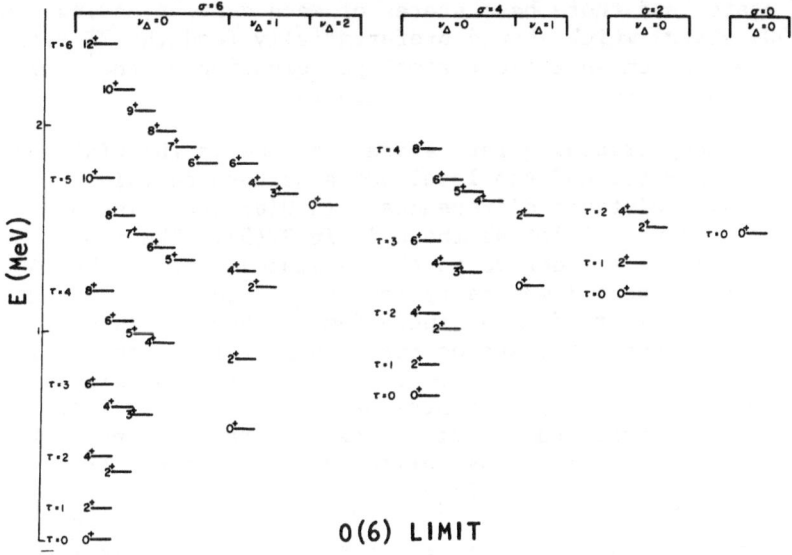

Fig. 1 A typical spectrum for a nucleus exhibiting the O(6) symmetry of the IBA. The energy levels are given by eq. 1 for N=6 with A=100 keV, B=30 keV and C=5 keV.

displaced, deformed γ-unstable oscillator[12] and the energies within a σ group follow the characteristic $\tau(\tau+3)$ law of that geometrical analogue of the O(6) symmetry. A third quantum number, ν_Δ, further subdivides the levels of the O(6) limit according to the number of zero coupled boson triplets.

The characteristic O(6) E2 selection rules are $\Delta\sigma=0$ and $\Delta\tau=\pm1$. The former arises from a detailed cancellation in the sum of terms arising in each matrix element and, as such, is rather easily broken by any perturbation to the ideal limit. The τ selection rule, however, results from the fundamental structure of the non-zero amplitudes for each O(6) state and is ultimately related to the more basic $\Delta n_d=\pm1$ selection rule for electromagnetic operators in any boson model. Breaking of the τ selection rule therefore requires the explicit introduction of a $\Delta n_d=\pm1$ changing perturbation (such as an interboson $\bar{Q}\cdot\bar{Q}$ interaction—see below). Through the use of group theoretic methods, the IBA provides analytic expressions for branching ratios of all transitions in the three limits. It is important to emphasize that the expressions that result are <u>totally independent of all parameters</u>, being characteristic of the limits themselves.

A characteristic signature of the O(6) limit, arising from the repeating sequences of states for each σ value and from the τ selection rule, is the appearance of $0^+-2^+-2^+$ sequences connected by E2 cascade transitions. Another signature is that the 0^+ states fall into two classes, those with high τ (such as the $(\sigma,\tau,\nu_\Delta)=(631)$ level) which should preferentially decay to the 2_2^+ level rather than the 2_1^+ state, and those base states of each σ group having $\tau=0$ (e.g., the (400) state) which should preferentially feed the 2_1^+ rather than the 2_2^+ level given an infinitesimal perturbation of the O(6) wave functions which breaks the σ selection rule.

While many branching ratios are the same in the O(6) and SU(5) or vibrator limits, and the level patterns have certain resemblances, there are also distinct differences. In O(6) there is no 0^+ member of a "two phonon" triplet as there is in SU(5). The $\tau=3$ 0^+ state is not this level (it decays to the 2_2^+ state), nor is the (400) 0^+ level which has no allowed decay routes. Further, all the states of the SU(5) symmetry form a single family whereas, in O(6), the σ quantum number separates states into distinct families. Thirdly, the level energies follow a $\tau(\tau+3)$ law instead of showing a proportionality to τ. Other differences arise from the explicit treatment of the finite dimensionality of the extra-core nucleons. This is exhibited, for example, in deviations of B(E2) values from the values given by the geometrical analogues of the IBA. (Thus, in SU(5), $B(E2:4_1^+ \to 2_1^+)/B(E2:2_1^+ \to 0_1^+) = 2(N-1)/N)$. In general, the effects of finite dimensionality are larger for smaller N and larger n_d. Since each O(6) wave function consists of several amplitudes for basis wave functions, each of different n_d, while the SU(5) wave functions are diagonal in a basis where n_d is a good quantum number, the expectation values of n_d in the two limits will differ and therefore also the size of the finite dimensionality effects. This aspect of the IBA is even more evident in the level energies. In SU(5), the eigenvalues are independent of N. In O(6), this is also true for the $\sigma=\sigma_{max}=N$ group but the energies of all higher energy (lower σ) groups explicitly (see eq. (1)) depend on N.

Finally, we note the particular suitability of the non-selective (n,γ) reaction for comprehensively testing the IBA since the model states encompass a broad variety with many degrees of freedom. More selective reactions access only a few of the predicted levels. For example, Coulomb excitation cannot excite any of the $\sigma<N$ states of the O(6) limit, even though they are collective, since the σ selection rule effectively isolates them. Similarly, the (t,p) reaction to 0^+ levels will populate the ground state and the base state of the $\sigma=N-2$ group, but not, for example, the first excited, (631), 0^+ state which involves a zero coupled triplet of bosons.

III. ^{196}Pt: AN O(6) NUCLEUS

From the (n,γ) (and other) studies mentioned earlier a complex level scheme for ^{196}Pt was developed consisting of all 0^+, 1^+, 2^+ levels below $\sim$2.5 MeV as well as a number of other levels. We extract from the full scheme[5] the decay properties for the positive parity levels below the pairing gap at $\sim$1.8 MeV. All levels have uniquely assigned J^π values and all γ-ray placements are considered firm. One interesting feature is the 2^+, 4^+ doublet near 800 keV, with the conspicuous absence of a close lying 0^+ state. This is a signature of the O(6) symmetry and, combined with the a priori expectation[13] that this limit might apply in this region encourages one to attempt such an interpretation. Several high lying states are depopulated by low energy transitions. These represent the largest decay matrix elements for these states and help establish familial relationships. Given these, along with the decay preferences for the 0^+ states, it is easy to assign O(6) quantum numbers to the empirical levels of ^{196}Pt. This is shown in Fig. 2 (see refs. 5 and 9 for detailed arguments). Note that the scheme involves three different σ groups and three characteristic $0^+-2^+-2^+$ sequences, two occurring rather high in energy.

Overall, the agreement with the predictions of the O(6) limit is remarkable. (There are clear discrepancies in energies but this is not unexpected: in a more refined IBA model, incorporating both proton and neutron bosons, it is found,[13] for nuclei near the O(6) limit, that the energies are significantly altered, and in such a direction as to agree markedly better with the empirical energies, while the E2 branching ratios are changed only slightly.) To summarize the agreement, up to $\sim$1.8 MeV, every low spin level predicted by the O(6) limit is observed, and, conversely, every observed level has an O(6) counterpart. Analogously for relative B(E2) values, all

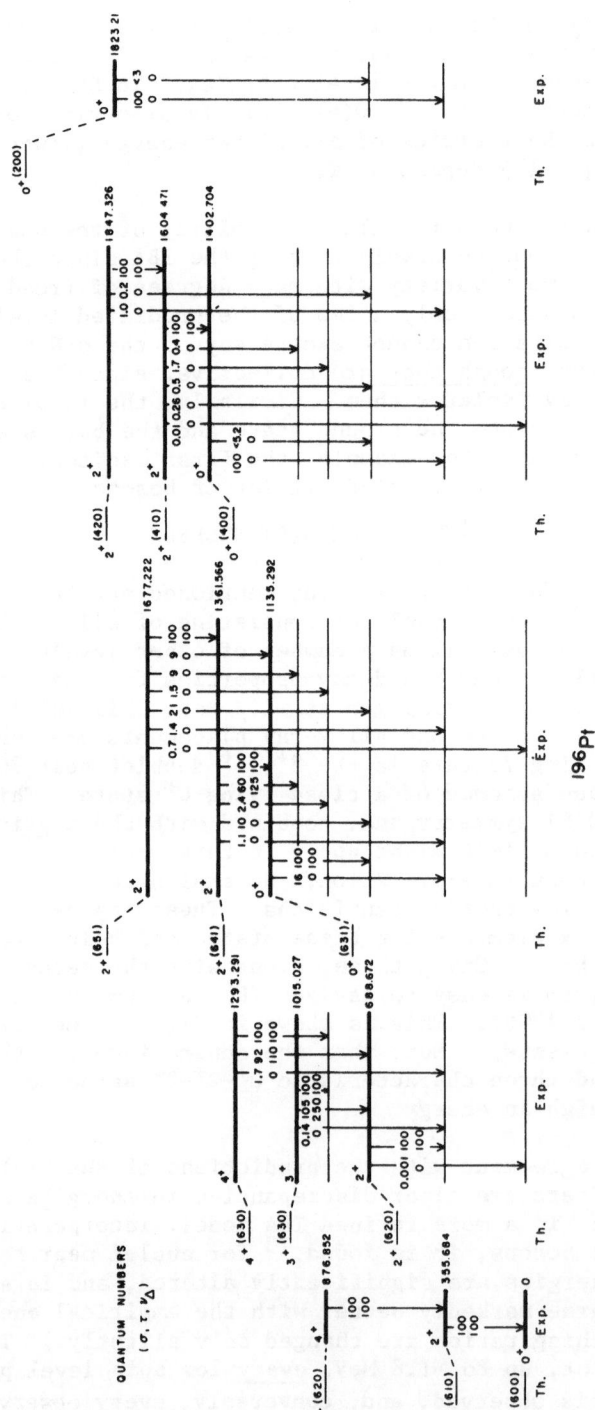

Fig. 2: Comparison of the positive parity levels of ^{196}Pt with the O(6) limit for N=6 and the parameters A = 185 keV, B = 43 keV, C = 23 keV. The O(6) quantum numbers $(\sigma, \tau, \nu_\Delta)$ are indicated for each theoretical level. For each level the upper (lower) row of numbers on the transition arrows give the measured (O(6) predicted) relative B(E2) values.

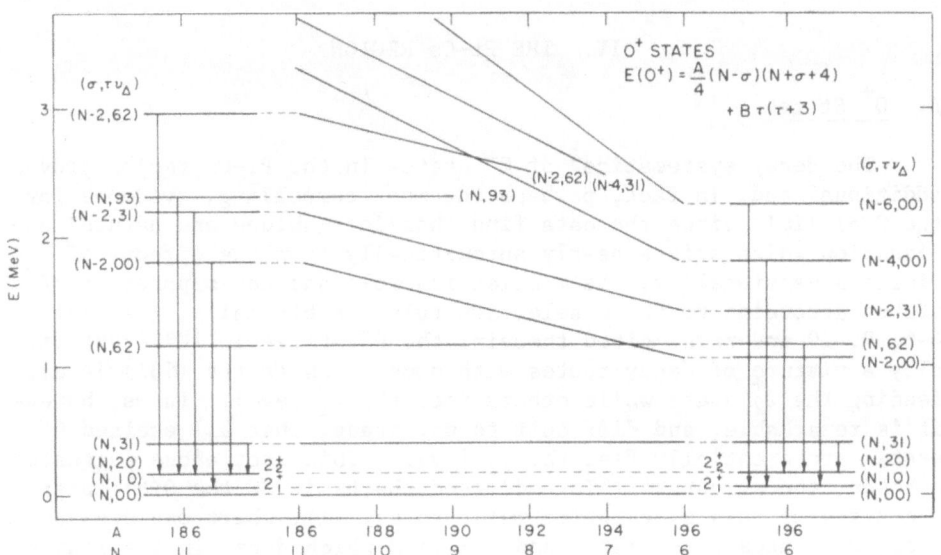

Fig. 3: N dependence of the decay properties of 0^+ states near the
0(6) limit. For N=6 and 11 (mass = 196 and 186), the level scheme
for 0^+ states is drawn at the right and left, respectively, with the
preferred E2 decay routes (to either the 2_1^+ or 2_2^+ level) shown: this
route is either that allowed in the 0(6) scheme or that involving the
least degree of forbiddeness in a slightly perturbed 0(6) scheme.

those predicted to be strong are strong and all predicted to be forbidden are weak or unobserved. Further, we reemphasize that these
branching ratio predictions are independent of any parameters. The
purity of the O(6) wave function is apparent from the fact that all
forbidden transitions are at least an order of magnitude or more
weaker than the allowed ones.

Thus, to reiterate, ^{196}Pt appears to provide an excellent empirical manifestation of the O(6) symmetry in nuclei. This, in turn,
gives strong support for the basic structure of the IBA in which this
symmetry is inherently contained.

IV. THE Pt-Os REGION

A. 0^+ States

The decay systematics[4] of 0^+ states in the Pt-Os region provides
additional and, in fact, perhaps the most compelling, evidence for
the O(6) limit since the data find therein a unique and natural explanation which arises nearly automatically from the effects of
finite dimensionality, the τ selection rule and the separation of
states according to the σ selection rule. Empirically, (see refs.
4-6, 9, 10 and refs. cited therein) the 0^+ states in ^{194,196}Pt display a mixture of decay routes with some, such as the (N31) level,
feeding the 2_2^+ state while others feed the 2_1^+ level. In Os, however,
it is remarkable, and difficult to undertand, that all excited 0^+
states preferentially feed the 2_2^+ level. This fact alone suggests[4]
a boson type structure. To undertand the basis of the O(6) interpretation of this systematics refer to eq. (1), where one sees that
those 0^+ states with high τ and σ=N (and which decay in a slightly
perturbed O(6) scheme to the 2_2^+ level rather than the 2_1^+ level) remain constant in energy as N changes while those with τ=0 and σ<N
(and which preferentially feed the 2_1^+ state) increase in energy with
N. As shown in Fig. 3, a N varies from 6 to 11 most of the latter
rise out of the low energy region. This, combined with the emergence, for N≥9, of new, high τ 0^+ states, implies that the dominant
decay of low lying 0^+ states for A∿188 in Os or Pt should be to
the 2_2^+ state while a mixture of decay routes should characterize
nuclei with A∿196.

B. The O(6) → Rotor Transitions

The recognition of the O(6) symmetry in ^{196}Pt now provides
a new touchstone for considering the empirical characteristics of
neighboring nuclei: that is, just as the recognition of the J (J+1)
law for the deformed rotor immediately gave meaning to the observed
small deviations therefrom in certain nuclei, one can now look at
the empirical properties of the Pt-Os nuclei, relative to this new

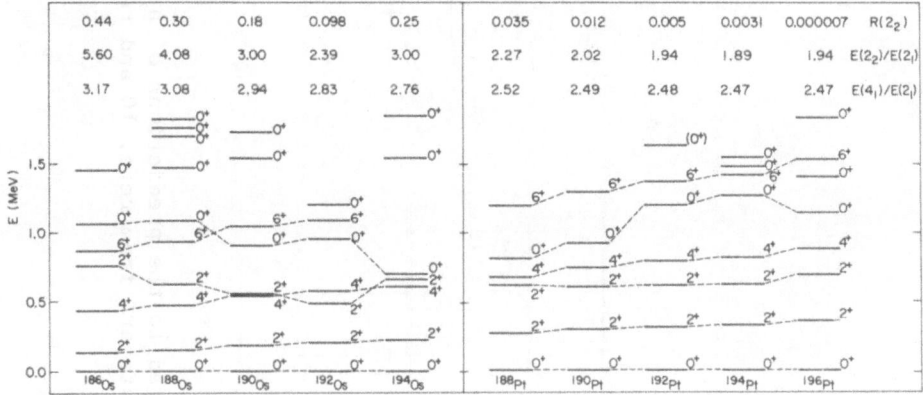

Fig. 4: Level systematics of the Pt-Os nuclei. $R(2_2)$ is defined as $B(E2:2_2 \rightarrow 0_1)/B(E2:2_2 \rightarrow 2_1)$. The data are described in refs. 10.

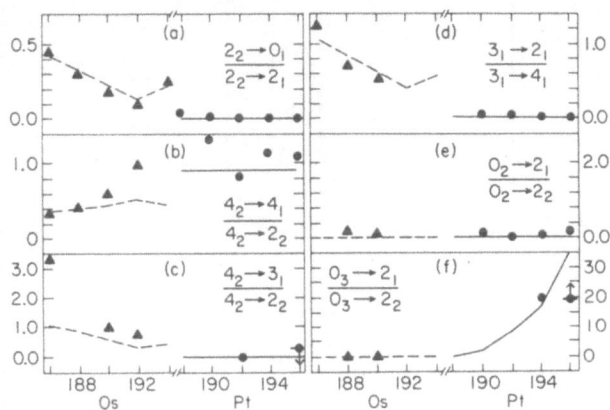

Fig. 5: E2 branching ratio systematics. The points are the empirical values (see refs. 10 and refs. cited therein). The solid and dashed lines connect the theoretical predictions. Note the radically different vertical scales, especially in boxes e and f.

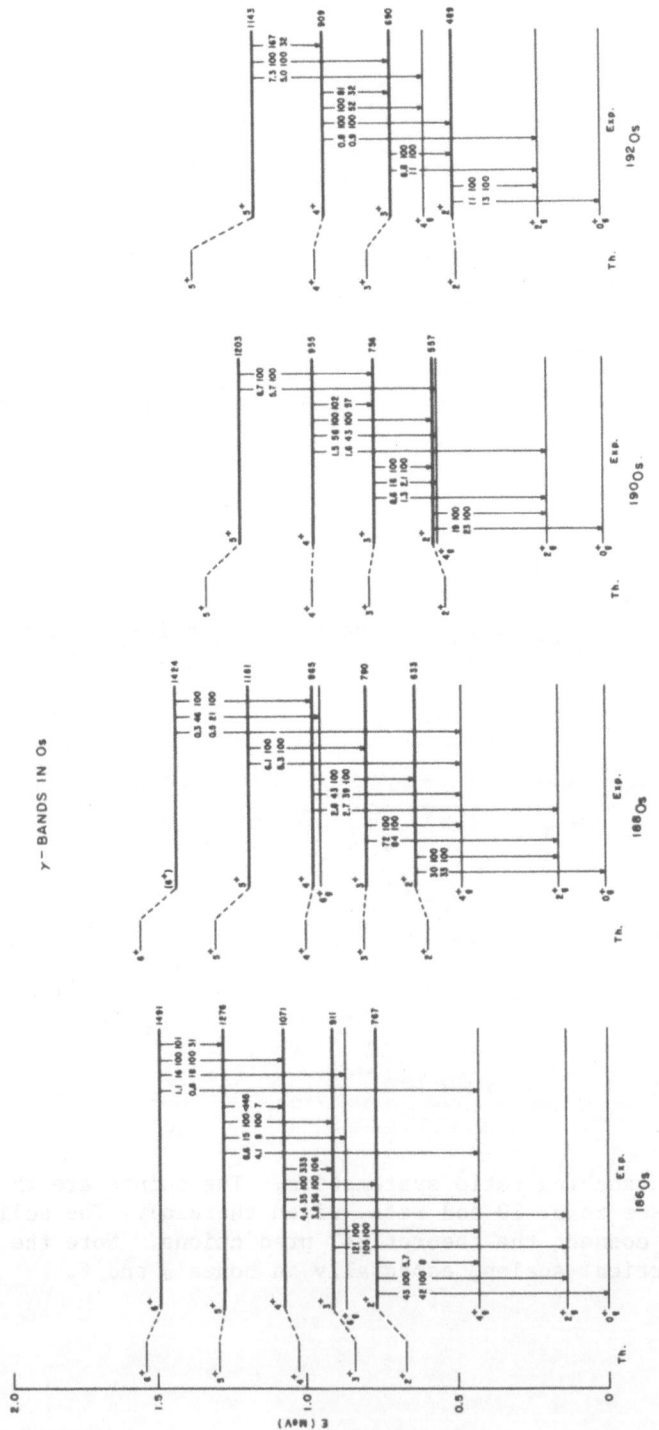

Fig. 6: Relative B(E2) values for quasi-γ-bands in Os compared to the predictions of a slightly perturbed O(6) scheme. The format is similar to Fig. 2. Data are from refs. 10 and refs. cited therein.

standard, to determine if these properties now disclose the operation of relatively simple perturbations to the O(6) scheme. Figure 4 summarizes a few of the relevant level energy and branching ratios systematics. It is clear that, for each element, as N decreases, the decreasing $E(2_1^+)$, increasing $E(4_1^+)/E(2_1^+)$, increasing branching ratio $R(2_2) = B(E2:2_2^+ \to 0^+)/B(E2:2_2^+ \to 2_1^+)$ and, in Os, the rapidly rising $E(2_2^+)$ all suggest[14] the approach of a region of deformation.

These observations suggest that perhaps one could interpret the region as initiating an O(6) → rotor transition. Such a transition has profound effects on a number of E2 branching ratios since many transitions that are allowed (forbidden) in O(6) become weak interband (enhanced intraband) transitions in the rotor, while branching ratios involving other transitions do not change significantly. Typically, as is evident from Fig. 1 and the τ selection rule, most O(6) levels have several allowed deexcitation routes, usually with comparable B(E2) values, while the allowed deexcitations in the rotor are usually only the one or two possible intraband transitions. Support for an O(6) → rotor interpretation of the Pt–Os region is provided by the branching ratios given in Table 1.

Table 1. Relative E2 transition probabilities in the O(6) and symmetric rotor models with empirical values (see refs. 10 and refs. cited therein) for nuclei in the Pt–Os region.

	$\dfrac{2_2 \to 0_1}{2_2 \to 2_1}$	$\dfrac{2_2 \to 2_1}{2_1 \to 0_1}$	$\dfrac{3_1 \to 2_1}{3_1 \to 4_1}$	$\dfrac{4_2 \to 4_1}{4_2 \to 3_1}$	$\dfrac{4_2 \to 3_1}{4_2 \to 2_2}$	$\dfrac{4_3 \to 3_1}{4_3 \to 2_2}$	$\dfrac{2_3 \to 3_1}{2_3 \to 0_2}$
O(6) N=6	0	1.31	0	∞	0	∞	1.25
196–Pt	0.000007	1.10	0.0014	>3.5	<0.31		0.60
190–Pt	0.012		0.038				0.47
192–Os	0.105	0.85		1.2	0.81	6.25	
188–Os	0.30	0.28	0.72			1.9	0.20
Rotor	0.70	small (∼0.02)	2.50	small (∼0.02)	2.23	0.56	small (∼0.02)

Given these encouraging trends, we attempted a series of schematic calculations, using the general IBA computer code[15] PHINT. This code diagonalizes the full IBA Hamiltonian[1-3] and thus allows calculations for nuclei deviating from the three limiting symmetries. In utilizing the code, we adopted a Hamiltonian of the schematic form $H=H(O(6))-\kappa Q \cdot Q$ in which the pure O(6) symmetry is allowed to be broken by a $\bar{Q} \cdot \bar{Q}$ interboson force that grows systematically with decreasing N and Z. (In the limit in which this force dominates one recovers the SU(3) limit.)

The parameters of eq. (1) and κ were obtained by rough fits of the level energies of the Pt-Os nuclei subject also to the contraints imposed by the branching ratio $R(2_2)$. The parameters were not fine tuned but were forced to very smoothly, usually being either linear or constant within each element. Furthermore, as shown in Table 2, for the E2 properties of these nuclei, essentially the only relevant parameter is the ratio κ/B which specifies the relative location of a nucleus between the O(6) and rotor limits. Not surprisingly, this ratio is small throughout this region but still nearly an order of magnitude larger in Os than in Pt. Note also that the effects of the changing boson number, N, are at least as significant as those due to changes in κ/B. The insensitivity of the branching ratios to the parameters A and C should not be surprising. A affects the relative location of the different σ groups and therefore, unless such groups are very close lying, it has little effect on the branching ratios within a group. The degeneracy breaking C term is diagonal in the O(6) Hamiltonian and thus will not affect branching ratios.

The results of the calculations consist primarily of a large number of E2 branching ratios and absolute B(E2) values. We can only discuss a few examples here. Figure 5 shows some branching ratios in the Pt and Os isotopes. Boxes a, c and d involve transitions in the numerators ($2_2^+ \rightarrow 0_1^+$, $4_2^+ \rightarrow 3_1^+$ and $3_1^+ \rightarrow 2_1^+$) that are τ forbidden in O(6) but enhanced or at least allowed in the rotor and comparable in magnitude to those in the respective denominators. Box b shows a case of two allowed O(6) transitions becoming inter- and intra-band transitions, respectively in the rotor. Boxes e and f show the decay systematics for the first and second excited 0^+ states discussed earlier. One sees that, in most cases, Os and Pt exhibit radically different behavior, that all the Pt branching ratios remain very close to the O(6) limit whereas significant deviations therefrom occur in Os in the direction of the rotor limit. (See Table 1 for some of the ratios in these limits.) In all cases the calculations nicely track the empirical results without, of course, matching the fine details.

The results for 0^+ states in Fig. 5 show that the qualitative arguments made earlier work equally well in the detailed calculations.(Comparable agreement is obtained for all but one of the other 0^+ states as well.) Thus the 0_2^+ level decays primarily to the 2_2^+ level throughout the region whereas the 0_3^+ level changes character entirely. In the heavy Pt isotopes, it is the (N-200) state and preferentially decays to the $\tau=1$ 2_1^+ level. For larger N, in ^{188,190}Os, its closest geometrical analogue would be a type of multi-γ-vibrational phonon excitation decaying to the 2_2^+ or 2_γ^+ state.

In addition to the examples shown in Fig. 5 the same calculations predict rates for hundreds of other transitions in this mass region. As two examples, we show in Figs. 6 and 7 the systematics of relative B(E2) values for γ-vibrational bands in Os and for all known

Table 2.   Parameter dependence of relative $B(E2:J_i \to J_f)$ values calculated in the IBA.[a]

$J_i \to J_f$	Params. as in footnote: $\kappa/B=0.116$	A = 170 keV $\kappa/B = 0.116$	B = 23.0 keV $\kappa/B = 0.109$	C = 18 keV $\kappa/B = 0.116$	$\kappa = 2.875$ keV $\kappa/B = 0.134$	B = 20.0 keV $\kappa = 2.325$ keV $\kappa/B = 0.116$	N = 10 $\kappa/B = 0.116$
$2_2 \to 0_1$	24	24	22	24	27	24	28
$2_2 \to 2_1$	100	100	100	100	100	100	100
$3_1 \to 2_1$	13	13	13	13	13	13	13
$3_1 \to 4_1$	21	21	22	21	19	21	18
$3_1 \to 2_2$	100	100	100	100	100	100	100
$4_2 \to 2_1$	1.7	1.7	1.5	1.7	2.0	1.7	2.1
$4_2 \to 4_1$	43	43	44	43	41	43	41
$4_2 \to 2_2$	100	100	100	100	100	100	100
$4_2 \to 3_1$	59	59	54	59	68	59	73
$4_3 \to 2_1$	0.16	0.16	0.16	0.16	0.17	0.16	0.16
$4_3 \to 4_1$	1.5	1.5	1.6	1.5	1.3	1.5	1.2
$4_3 \to 6_1$	0.96	0.96	1.00	0.96	0.81	0.96	0.75
$4_3 \to 2_2$	16.5	16.4	14.6	16.5	21.4	16.5	23.4
$4_3 \to 3_1$	88	88	90	88	86	88	83
$4_3 \to 4_2$	100	100	100	100	100	100	100
$0_2 \to 2_1$	1.1	1.1	1.2	1.1	0.92	1.1	1.0
$0_2 \to 2_2$	100	100	100	100	100	100	100

a)   Cols. 3-8 give the calculated $B(E2)$ values for N=9 for various parameter changes relative to the values A=150, B=21.5, C=16, κ=2.5 keV of col. 2. Only the parameter changed with respect to col. 2 is given. In col. 7 both B and κ are changed while retaining the same ratio κ/B as in col. 2. The last column presents the values for N=10 with the parameters of col. 2.

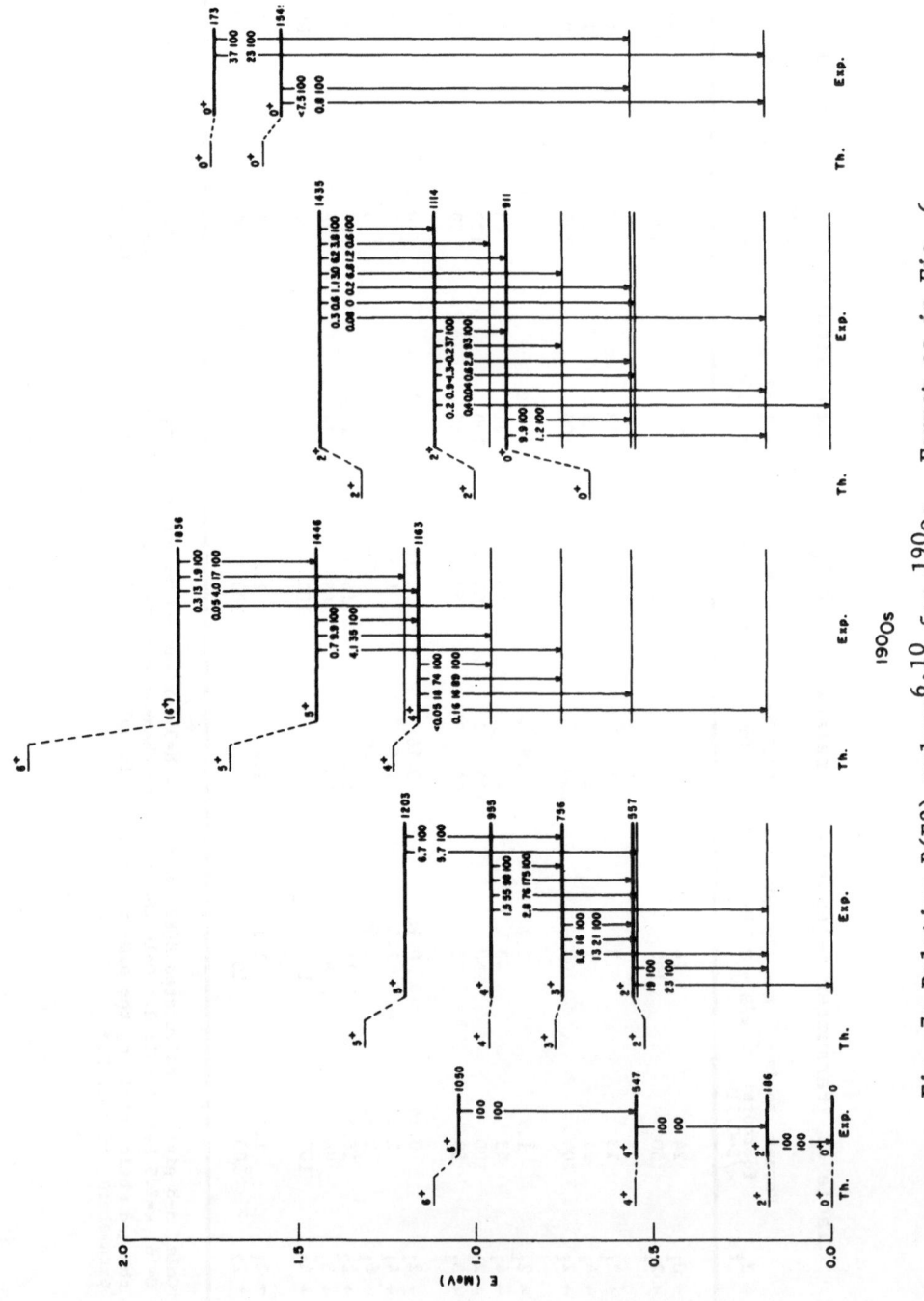

Fig. 7: Relative B(E2) values[6,10] for ^{190}Os. Format as in Fig. 6.

collective levels in one specific nucleus, ^{190}Os. Note, in Fig. 6, the systematic changes across a series of isotopes, changes which are very closely reflected in nearly all cases in the calculated B(E2) values. It is interesting to observe in Fig. 7 that, by ^{190}Os, definite characteristics of deformed nuclei have emerged such as the existence of well defined γ and K=4 rotational bands (note the strong $4^+_\gamma \rightarrow 3^+_\gamma$ transition) at the same time that remnants of the O(6) scheme persist, such as the $0^+-2^+-2^+$ sequence based on the 0^+ state at 911 keV.

While these branching ratios provide an extensive test of the model, one is also particularly interested in the predictions for <u>absolute</u> B(E2) values which can be calculated from the wave functions obtained earlier. Some of the results are shown in Fig. 8, which also includes the microscopic calculations of Kumar and Baranger,[18] heretofore generally considered the best for this region. In every case, as well as a number of others not shown, the agreement is excellent and in fact, always better than that of ref. 18.

Since these calculations were performed we have become aware of recent measurements[19] and extractions[20] of additional B(E2) values

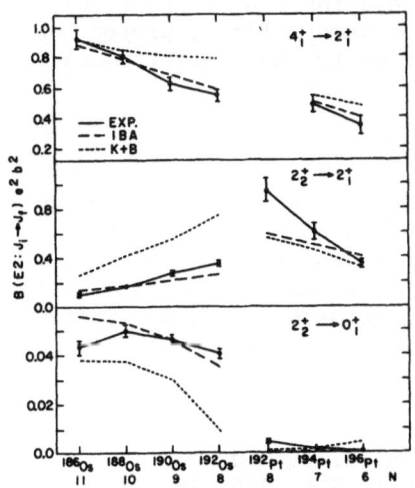

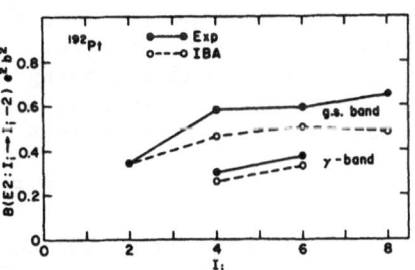

Fig. 8: Absolute B(E2) values for Os and Pt. The measured values[16,17] are connected by solid lines. The IBA predictions and those of Kumar and Baranger[18] are indicated by dashed lines. All B(E2) values here and in Fig. 9 are normalized to the experimental value for the $2^+_1 \rightarrow 0^+_1$ transition.

Fig. 9: B(E2) values for ^{192}Pt from ref. 19 compared to the IBA predictions using the same calculations as for the previous comparisons. The calculated B(E2) values are normalized to the B(E2: $2^+_1 \rightarrow 0^+_1$) adopted in ref. 19.

for the quasi-γ bands in Pt. The existing calculations of course already contain the IBA predictions. Once again, the agreement is excellent. One example, for ^{192}Pt, is illustrated in Fig. 9.

VI. SUMMARY

We feel that the results summarized above provide substantial support for the existence of the O(6) symmetry, for an O(6) $\rightarrow$ rotor transtiion in the Pt-Os nuclei and, overall, for the basic validity of the IBA scheme. In general an anormous amount of empirical data is rather accurately reflected in a simple set of calculations dependent on basically a single parameter. In fact, many of the correct predictions, however, result from the fundamental structure of the IBA including its explicit treatment of finite dimensionality, and are indeed essentially parameter <u>independent</u>. These results also point to a number of key tests of the IBA that should be attempted such as the search for other O(6) nuclei (e.g., ^{130,132}Xe, ^{134,136}Ba), more comprehensive tests of nuclei approximating the SU(3) and SU(5) limits, tests of predictions for two nucleon transfer cross sections for which elementary selection rules may be easily stated in the three IBA limits and, finally, the search for discrepancies between the empirical data and the IBA, in particular discrepancies that might reveal where a more refined, proton boson-neutron boson model, is markedly superior to the simpler model discussed above.

ACKNOWLEDGEMENTS

This work was performed in close and essential collaboration with Dr. J. A. Cizewski with whom it was a pleasure to work. I am deeply grateful to Prof. F. Iachello for considerable and continuing help in understanding the IBA model and in applying it to the Os-Pt region. Finally, I acknowledge with gratitude fruitful discussions with Professors A. Arima and I. Talmi.

REFERENCES

*   Work supported by US DOE Contract No. EY-76-C-02-0016.
1. F. Iachello and A. Arima, Phys. Lett. 53B, 309 (1974);
    F. Iachello, Nukleonika 22, 107 (1977).
2. A. Arima and F. Iachello, Ann. Phys. (N.Y.) 99, 253 (1976);
    A. Arima and F. Iachello, Ann. Phys. (N.Y.), to be published,
    (preprint, KVI-105 Groningen, Holland, 1977); A. Arima and
    F. Iachello, Phys. Rev. Lett. 35, 1069 (1975); O. Scholten,
    F. Iachello and A. Arima, Ann. Phys. (N.Y.), to be published
    (preprint, KVI-126, Groningen, Holland, 1978).
3. A. Arima and F. Iachello, Phys. Rev. Lett. 40, 385 (1978).
4. M. R. Macphail, R. F. Casten and W. R. Kane, Phys. Lett. 59B,
    435 (1975).

5. J. A. Cizewski, R. F. Casten, M. R. Macphail, G. J. Smith,
    M. L. Stelts, W. R. Kane, H. G. Börner and W. F. Davidson, to
    be published.
6. R. F. Casten, M. R. Macphail, W. R. Kane, D. Breitig, Klaus
    Schreckenbach and J. A. Cizewski, Nucl. Phys., to be published.
7. R. F. Casten, H. G. Börner, J. A. Pinston and W. F. Davidson,
    Nucl. Phys., to be published.
8. R. F. Casten, A. I. Namenson, W. F. Davidson, D. D. Warner and
    H. G. Börner, Phys. Lett. 76B, 280 (1978).
9. J. A. Cizewski and R. F. Casten, Phys. Rev. Lett. 40, 167 (1978).
10. R. F. Casten and J. A. Cizewski, Phys. Lett. 79B, 5 (1978) and
    Nucl. Phys. A309, 477 (1978).
11. R. F. Casten, contribution to the Third International Symposium
    on Neutron Capture Gamma-ray Spectroscopy and Related Topics,
    September 18-22, 1978, Brookhaven Nat'l. Laboratory (Plenum Press)
12. L. Wilets and M. Jean, Phys. Rev. 102, 788 (1956).
13. A. Arima, T. Ohtsuka, F. Iachello and I. Talmi, Phys. Lett.
    66B, 205 (1977); T. Ohtsuka, A. Arima, F. Iachello and I. Talmi,
    preprint, KVI-125, 1978; F. Iachello, private communication.
14. K. Kumar, Phys. Rev. C1, 369 (1970).
15. Computer Code PHINT, written by O. Scholten, KVI Groningen,
    Holland.
16. I. Berkes, R. Rougny, Michele Meyer-Levy, R. Chery, J. Daniere,
    G. Lhersonneau, and A. Troncy, Phys. Rev. C6, 1098 (1973).
17. R. F. Casten, J. S. Greenberg, S. H. Sie, G. A. Burginyon,
    and D. A. Bromley, Phys. Rev. 187, 1532 (1969); W. T. Milner,
    F. K. McGowan, R. L. Robinson, P. H. Stelson and R. O. Sayer,
    Nucl. Phys. A177, 1 (1971).
18. K. Kumar and M. Baranger, Nucl. Phys. A122, 273 (1968).
19. C. Roulet, H. Sergolle, P. Hubert and Thomas Lindblad, Physica
    Scripta 17, 487 (1978).
20. K. Stelzes, K. Rauch. Th. W. Elze, Ch.E. Gould, J. Idzko,
    G. E. Mitchell, H. P. Nottrodt, R. Zoller, H. J. Wollersheim
    and H. Emling, Phys. Lett. 70B, 297 (1977).

COLLECTIVE PROPERTIES OF NUCLEAR STATES IN THE A∼150
TRANSITIONAL REGION AND THE INTERACTING BOSON MODEL

Ziemowid Sujkowski

Department of Physics /IA/
Institute for Nuclear Research
05-400 Świerk near Warsaw
Poland

A consistent description of collective properties of nuclei having no distinct geometrical symmetries is an obvious challenge to any nuclear model. One of the attractive features of the Interacting Boson Model [1,2] is that it attempts to do just that in a relatively simple and physically transparent way. It is thus of general interest to test this description by carrying out numerical calculations for a number of observables for e.g. a set of isotopes spanning the "vibrational" - to - "rotational" transition region. In the terms of the IBA the calculations correspond to solving a simplified Hamiltonian chosen for intermediate situations between the vibrational, SU /5/, and rotational, SU /3/, limiting cases. Partial results of such an analysis for the samarium isotopes have been presented in this talk and compared with the experimental data. Since the time of the Erice conference, however, a much more complete analysis of these nuclei has been published [3]. Properties such as the energy spectra, electromagnetic transition strengths, multipole moments and nuclear radii have been calculated with several simplifying assumptions, and with the aim to reproduce the trends in the data rather than the detail structures in each of the nuclei studied. Rather than repeating here the results of this or of similar analyses, the reader is referred to ref. 3.

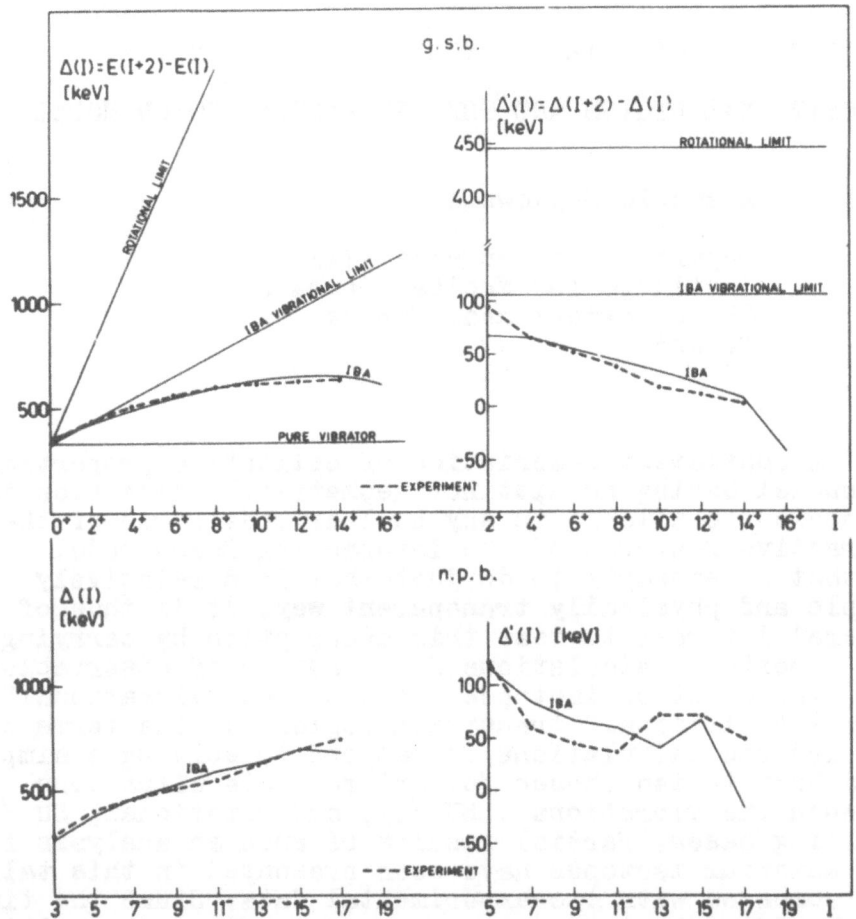

Fig. 1. First and second differences of the experimental and calculated level energies of g.s.b. and n.p.b. plotted as functions of the level spins.

A somewhat more detailed description of collective
properties of a particular transitional nucleus can
also be attained in the framework of the IBA at still
moderate computational costs. This has recently been
demonstrated [4] for the "classical" transitional nucleus
^{150}Sm. Here we shall quote only partial results of this
work concerning the energies of the ground state band
and of the negative parity band. The occurrence of the
latter band is a salient feature of nuclei in this mass
region. Assuming that the states in this band result
from the coupling of an octupole collective excitation
to the ground state band and retaining only two terms
in the octupole boson – quadrupole boson interaction
/the L·L term and the quadrupole term Q·Q / one can
relatively easily solve the corresponding total Hamiltonian. Fig. 1 shows the comparison of the calculated
energies with experiment for the ground state band and
the negative parity band. In order to dramatize the
correspondence between the calculated and the empirical
values, the first and the second differences of the
level energies are plotted rather than the total values.
Results expected for the limiting situations are also
shown. It is worth mentioning that the same calculation
yielded also the B/E1/ and B/E2/ values for intra- and
inter–band transitions. The B/E1//B/E2/ branching ratios
for transitions depopulating states in the n.p.b. were
found in very good agreement with experiment, while
those for transitions depopulating the g.s.b. were
generally overestimated. It was concluded in [4] that
it was possible to give a consistent description of
the ^{150}Sm nucleus in the framework of the Interacting
Boson Model with few adjustable parameters. It remains
to hope that with the Interacting Boson Model progressing
towards its higher degrees of sofistication the freedom
in selecting those parameters will be more and more
limited.

References

1. F. Iachello and A. Arima, Phys. Rev. Lett. 35 (1975)
   1069.
2. A. Arima and F. Iachello, Ann. of Phys. 99 (1976) 253.
3. O. Scholten, F. Iachello and A. Arima, Ann. of Phys.
   115 (1978).
4. Z. Sujkowski, D. Chmielewska, M. J. A. de Voigt,
   J. F. W. Jansen and O. Scholten, Nucl. Phys. A291
   (1977) 365 .

COLLECTIVE STATES IN SOME TRANSITION NUCLEI WITH $28 \leq Z, N \leq 50$

A. Gelberg and U. Kaup

Institut für Kernphysik der Universität zu Köln
Köln, West Germany[+]

ABSTRACT

Some general properties of collective states in nuclei with $28 \leq Z, N \leq 50$ are discussed. Energy levels and B(E2) values in even Kr isotopes were calculated by means of IBA. The description of odd nuclei in the same region by both IBA and the rotor + particle models is discussed.

Transition nuclei with $28 \leq Z, N \leq 50$ have been extensively studied during the last few years; in particular isotopes of Ge, As, Se, Br, Kr, and Br have been investigated mainly by means of in-beam gamma-ray spectroscopy[1-11].

The systematics of excitation energy and the large values of B(E2) show that collecitve excitations dominate in an energy range up to a few MeV.

Let us analyze several cases of both even-even and even-odd nuclei:

1. <u>EVEN-EVEN NUCLEI</u>

Collective features are characterized by a rather smooth dependence on either Z or N. Let us have a look e.g. at the energies of the first 2^+ state in even-even nuclei in this mass region (fig. 1). One can see that for $Z > 32$, each series of isotopes shows an increase of E_2+ when going from the middle of the shell towards $N = 50$.

[+] Supported by the Bundesminsterium für Forschung und Technik

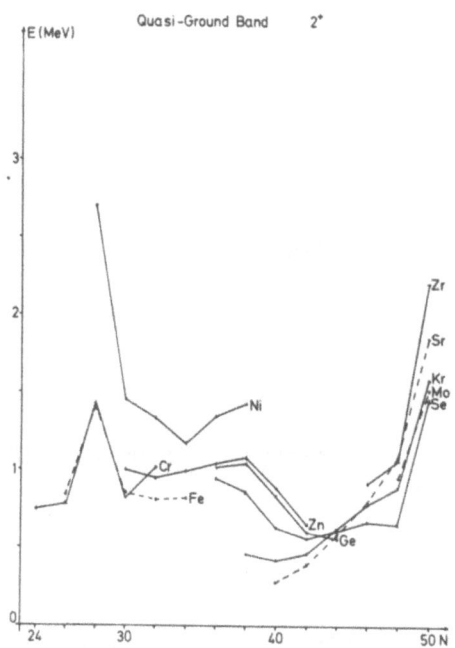

Fig. 1: Energies of 2^+ levels vs. neutron number

As an example we will discuss a case where there are enough experimental data, viz. that of Krypton. The positive parity states of Kr isotopes are represented in fig. 2. A smooth transition from a rotational pattern (e.g. ^{76}Kr) to a vibrational one (e.g. ^{84}Kr) can be observed. This is the type of transition which should be well described by IBA. The 0_2^+ state, conventionally described as the beta band head, behaves in a quite peculiar way, e.g. its energy dependence on N is nearly linear. This state cannot be reproduced by the present calculation and we can only say that it cannot be described in a simple way in terms of d- and s-bosons[12].

The positive parity states in ^{80}Kr can be easily reproduced even by the simplified IBA-1 model[13]. The situation is more complex in ^{78}Kr and the IBA-2[14] hamiltonian was used:

$$H = \epsilon(n_{d_\pi} + n_{d_\nu}) - K\, Q_\pi \cdot Q_\nu + V$$

where

$$Q_\pi = \left[(d^+ s + s^+ d)_\pi + \chi_\pi (d^+ d)_\pi \right]^{(2)}$$
$$Q_\nu = \left[(d^+ s + s^+ d)_\nu + \chi_\nu (d^+ d)_\nu \right]^{(2)}$$

and V is the Majorana term; a fit of ^{78}Kr, together with the experimental levels[15] is shown in fig. 3; the agreement is satisfactory up to the 10^+ yrast state where backbending occurs.

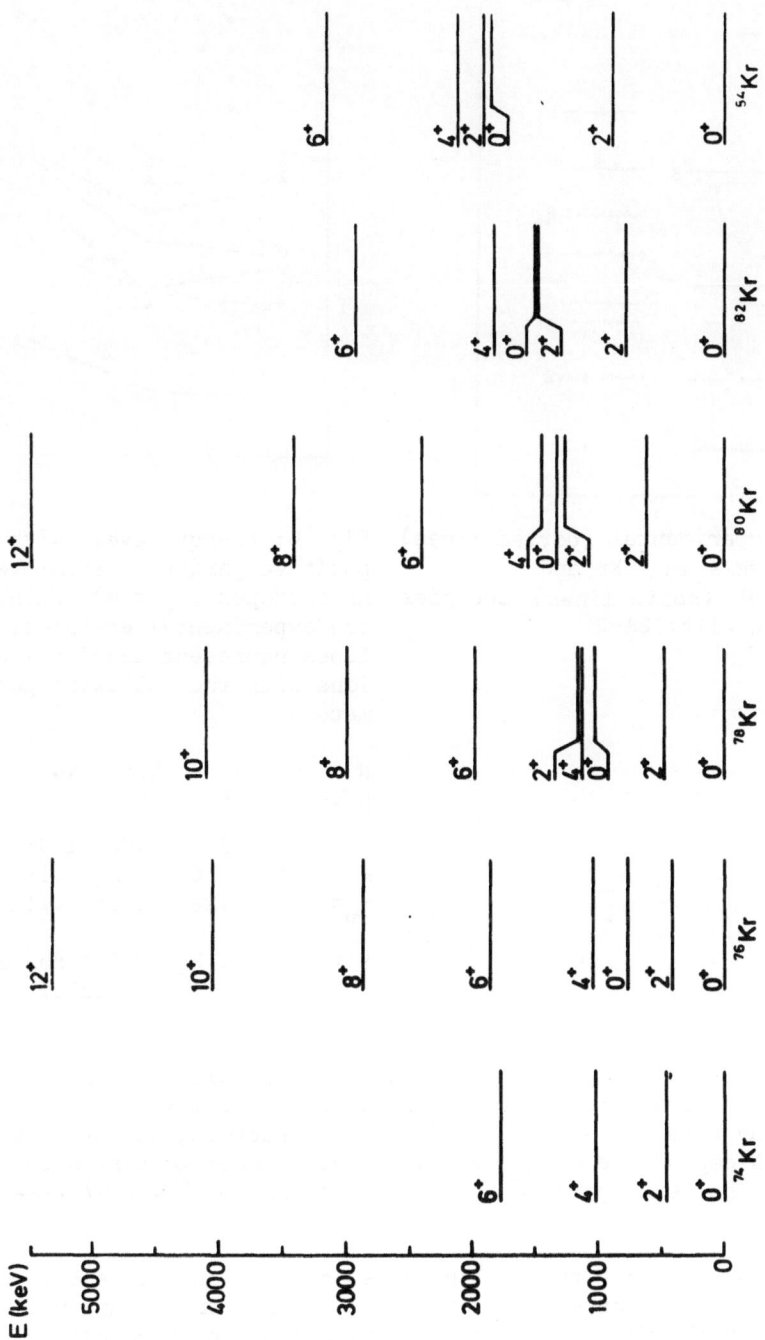

Fig. 2: Energy levels with positive parity in Kr isotopes (from refs. 7, 8, 9, 15, 20)

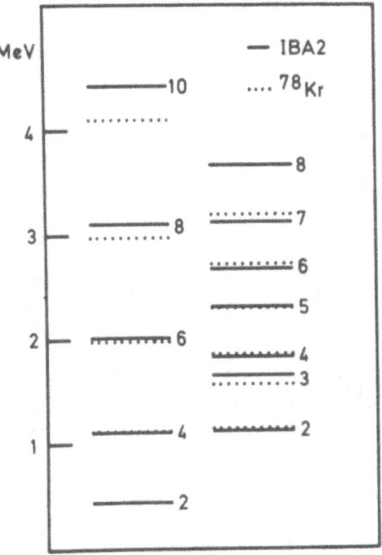

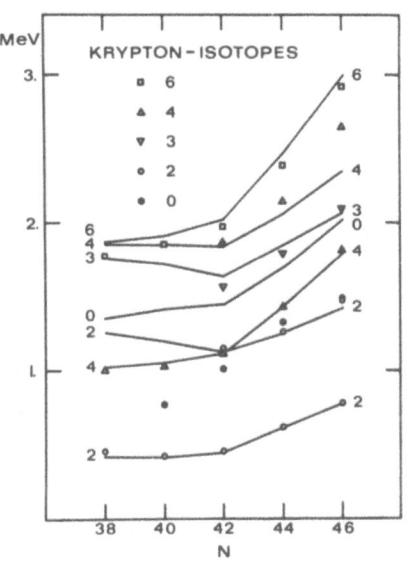

Fig. 3: Experimental (dotted lines) energy levels in ^{78}Kr and theoretical (solid lines) energies calculated with IBA-2.

Fig. 4: Energy levels with positive parity in even-even Kr isotopes (N_π = 4). Points are experimental energies. The lines represent IBA-2 calculations with the following parameters:

N =	38	40	42	44	46
N_ν =	5	5	4	3	2
ε =	1	1	.96	1.05	1.15
κ =	.15	.16	.18	.18	.19
χ_ν =	.07	.28	.495	.71	.92

χ_π = -1.127, F_κ = 0.1 for all nuclei

    This has encouraged us to try to describe all even Kr isotopes by means of a consistent set of constants. We kept χ_π constant, as well as the parameters of the Majorana interaction, and we assumed a linear N-dependence of χ_ν; κ and ε were allowed to vary smoothly (fig. 4). The level pattern is similar to that of the O(6) limit for N $\geq$ 42.

    The result shows that, although some individual levels in each nucleus are not exactly reproduced, the general trends are well described by the model. The use of separate π- and ν-bosons is appropriate, since Kr has 8 protons outside the closed shell Z = 28. while the neutrons are present as holes.

It is interesting to test the model on the behaviour of the reduced E2 transition probabilities along the yrast band. The Arima-Iachello model predicts deviations of the B(E2) value[13] in comparison with the leading order rotational formula[16]; this should happen mainly at high spin, owing to the finite number of bosons which can be built out of the particles belonging to the active shell. B(E2) values for ^{80}Kr, normalized to the $2^+ \rightarrow 0^+$ transition have been calculated using the wave functions which give the best fit for energy levels; a comparison between experiments[2,8] and several theoretical predictions is shown in fig. 5. Although the error bars are rather large and lifetimes have been measured only up to the 10^+ state, the data suggest a deviation from the rotational formula in the direction predicted by IBA. Preliminary data on E2 transitions in ^{78}Kr also show deviations[15].

The study of E2 transition probabilities between high-spin states constitutes a sensitive test for IBA. There are other suitable candidates, e.g. ^{80,82,84}Sr, ^{100}Ru, ^{100,102}Pd, etc. Since the admixture of a two-quasiparticle state can give rise to a similar effect, nuclei displaying strong backbending should be avoided.

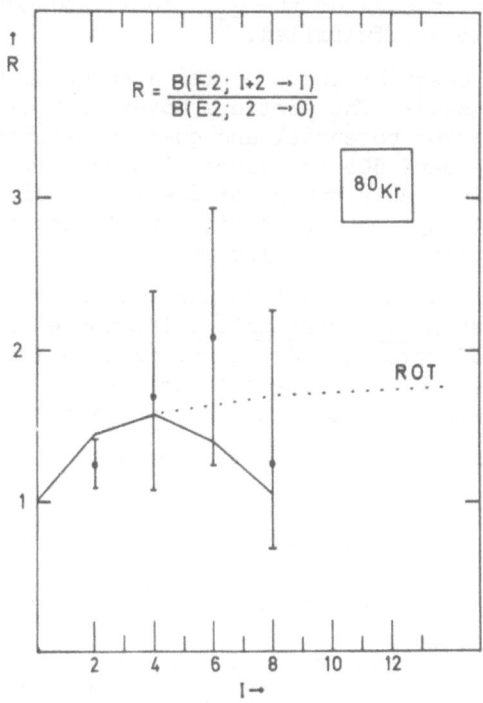

Fig. 5: B(E2) values in ^{80}Kr, normalized to the $2^+ \rightarrow 0^+_{g.s}$ transition. The data were compared with the predictions of the IBA-2 model (solid line) and the rigid rotor values (dashed line)

2. EVEN-ODD NUCLEI

Although the level schemes of odd nuclei are more complex, we can identify the members of collective bands; inside the bands the E2 transitions are strong, while interband transitions are much weaker. The following features are characteristic for odd nuclei in this mass region[1-4,6,11].

In odd-proton nuclei, like ^{69,71,73}As or ^{81,83,85}Rb, one can see positive parity, stretched $\Delta I = 2$ bands, with energy intervals roughly equal to those in the even-even core (decoupled bands). Negative parity bands in the same nuclei display strong staggering. This behaviour suggests a rotational interpretation with strong Coriolis coupling; the deformation should be $\beta \sim 0.2$.

The situation is different in neutron-odd nuclei such as odd Se and Kr isotopes. A correspondence between the levels of the positive parity band and those of the core is manifest, but the energy intervals are no longer equal; the resemblence to decoupled bands is only qualitative. As regards the negative parity bands they look very much like rotational sequences with little staggering. This behaviour suggests a larger deformation $\beta \sim 0.3$, which weakens the influence of the Coriolis coupling. Moreover, the Fermi level is pushed towards the middle of the $g_{9/2}$ shell and conditions for decoupling are no longer fulfilled.

In order to describe these nuclei a rotor plus particle Hamiltonian was used[17]. The particle moves in a deformed axially symmetric Woods-Saxon potential and quasi particles are generated by means of a standard BCS procedure. The collective rotational Hamiltonian with variable moment of inertia is diagonalized on this basis. In order to obtain a good fit for all positive parity states, triaxial deformation has to be assumed[18].

Since we deal with transition nuclei, the concept of "deformation" should be taken cum grano salis. The nuclei investigated are certainly no rigid rotors, and the use of a rotational basis represents only one possibility. The interacting boson model should again be able to describe such an intermediate situation.

Steps have been taken in this direction. A core-particle interaction Hamiltonian of the form [19]

$$H_{int} = \alpha \left[(d^{+}s + s^{+}d) - \frac{\sqrt{7}}{2} (d^{+}d) \right]^{(2)} \cdot (a_{j}^{+}a_{j})^{(2)}$$
$$+ \beta^{2}(d a_{j}^{+})^{(j)} \cdot (a_{j}d^{+})^{(j)}$$

has been used; $a_{j}^{+}(a_{j})$ is the creation (annihilation) operator for the particle with angular momentum j. This Hamiltonian is suitable only for nuclei where the particle outside the core occupies a single state j. The results for the positive parity levels in ^{81}Rb (^{80}Kr core) are shown in fig. 6, the agreement with experiment is excellent.

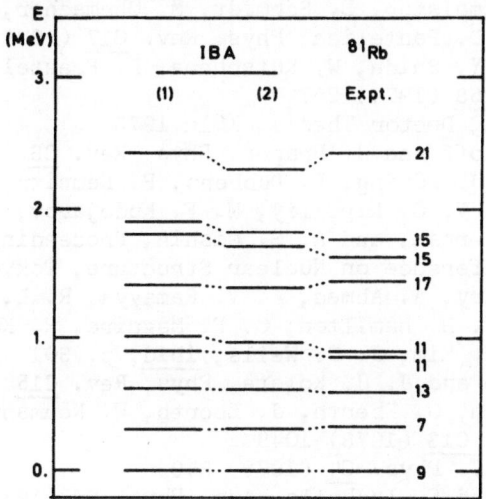

Fig. 6: Theoretical and experimental levels in ^{81}Rb. Two different cores were used: in case (1) the ^{80}Kr core was fitted by means of a pure SU(5) Hamiltonian, while case (2) represents a mixture of the SU(5) and O(6) limits. The two parameters describing the inter-action of the fermion with the core bosons have the same values α = 0.3 and β = 0.76 in both calculations.

The case of the negative parity levels requires a more elaborate calculation, since at least three particle states are mixed. Let us hope that the theory will be developed so as to allow the handling of such complicated cases without introducing a large number of free parameters.

The authors would like to express their gratitude to Prof. F. Iachello and Dr. O. Scholten for numerious stimulating discussions and for communication of programs and unpublished results.

REFERENCES

1. P. v. Brentano, B. Heits, C. Protop, in "Problems of vibrational nuclei", ed. G. Alaga, V. Paar and L. Sips, North Holland, Amsterdam 1975, page 155
2. H. G. Friederichs, A. Gelberg, B. Heits, K. P. Lieb, M. Uhrmacher, K. O. Zell and P. v. Brentano, Phys. Rev. Lett. 34 (1975) 745; H. G. Friederichs, A. Gelberg, B. Heits, K. O. Zell and P. v. Brentano, Phys. Rev. C13 (1976) 2247
3. B. Heits, H. G. Friederichs, A. Gelberg, K. P. Lieb, A. Peregro, R. Rascher, K. O. Zell and P. v. Brentano, Phys. Letters 61B (1976) 33; B. Heits, H. G. Friederichs, A. Rademacher, K. O. Zell and P. v. Brentano, Phys. Rev. C15 (1977) 1742

4. H. P. Hellmeister, E. Schmidt, M. Uhrmacher, R. Rascher, K. P. Lieb, and D. Pantelica, Phys. Rev. C17 (1978) 2113

5. E. Nolte, Y. Shida, W. Kutschera, R. Prestele, and H. Morinaga, Z. Phys. 268 (1974) 267

6. K. O. Zell, Doctor Thesis, Köln 1978

7. W. G. Wyckoff and J. Draper, Phys. Rev. C8 (1973) 796

8. L. Funke, J. Döring, F. Dubbers, P. Kemnitz, H. Strusny, E. Will, G. Winter, V. G. Kiptiliy, M. F. Kudojarov, I. Kh. Lemberg, A. A. Pasternak, and A. S. Mishin, Proceedings of the International Conference on Nuclear Structure, Tokyo 1977, p. 300

9. D. L. Sastry, A. Ahmed, A. V. Ramayya, R. L. Robinson, R. B. Piercey, J. H. Hamilton, C. F. Maguire, H. Kawakami, A. P. de Lima, H. J. Kim, J. C. Wells, ibid, p. 301

10. K. P. Lieb and J. J. Kolata, Phys. Rev. C15 (1977) 939

11. L. Cleemann, U. Eberth, J. Eberth, W. Neumann, and V. Zobel, Phys. Rev. C18 (1978) 1049

12. K. Kumar, J. Phys. G4 (1978) 849

13. A. Arima and F. Iachello, Ann. Phys. 99 (1976) 99, and O. Scholten (private communication)

14. T. Otsuka, A. Arima, F. Iachello, and J. Talmi, Phys. Lett. 76B (1978) 139

15. H. P. Hellmeister, J. Keinonen, K. P. Lieb, R. Rascher, R. Ballini, J. Delaunay, and H. Dumont, Contribution to the International Conference on Heavy Ion Collisions, Canberra (1978)

16. A. Bohr and B. Mottelson, Nuclear Structure, vol. II (W. A. Benjamin, Reading, Mass., 1975)

17. U. Kaup, A. Gelberg, and P. v. Brentano, (to be published)

18. H. Toki and A. Faessler, Z. Physik A276 (1976) 35

19. O. Scholten, private communication

20. J. Lin, A. C. Reuter, J. Wells, H. K. Carter, R. B. Piercey, A. V. Ramayya, J. H. Hamilton, R. L. Robinson, and H. J. Kim, BAPS 23 (1978) 574; R. B. Piercey, A. V. Ramayya, J. H. Hamilton, C. F. Maguire, R. L. Robinson, H. J. Kim, and J. Wells, BAPS 23 (1978) 574; R. L. Robinson, H. J. Kim, R. O. Sayer, R. D. Piercey, A. V. Ramayya, J. H. Hamilton, and J. C. Wells, BAPS 22 (1977) 1027

SYSTEMATICS OF EXPERIMENTAL PROPERTIES OF LOW-LYING STATES IN
EVEN Ru AND Pd NUCLEI

Lennart Hasselgren

Institute of Physics, University of Uppsala

Box 530, S-751 21 Uppsala, Sweden

It is well-known that the stable isotopes of Ru and Pd belong to a transitional region between a spherical and deformed shape of the potential surface. This is clearly reflected in the energy systematics of the positive parity states presented in fig. 1. In this figure only states with highly reliable spin assignments are given. Thus a number of states with positive parity are omitted especially in the energy region 1-2 MeV.

Nuclei in this mass region have for many years been interpreted to have an anharmonic vibrational stucture and fig. 1 shows that at N=56-58 a two-phonon multiplett seems to exist. Collective features are developed earlier as function of the neutron-number the more proton holes we have outside the closed shell. This is seen for the multiplett of states, 0_2^+, 2_2^+ and 4_1^+. The two-phonon multiplett can be followed from about N=60 for Cd nuclei (not shown), for Pd nuclei from about N=58 and for Ru nuclei from N=56. It is also worth noting that this anharmonic vibrational character seems to be broken quicker for Ru isotopes than for Pd and Cd isotopes. The second 0^+ state, 0_2^+, in the case of ^{108}Ru is even higher in energy than the first 3^+ state, 3_1^+, and the few states known indicate a transition towards a more rotational structure with γ-instability. The experimental situation for Pd isotopes does not give a clear indication of a raise in energy of 0_2^+ when passing the middle of the neutron shell, since the state at 928 keV in ^{112}Pd[1] has not been given any spin assignment and could thus be a 0^+ states.

For a more direct comparison with simple model predictions a few energy ratios are shown in fig. 2. Energy ratios for the yrast

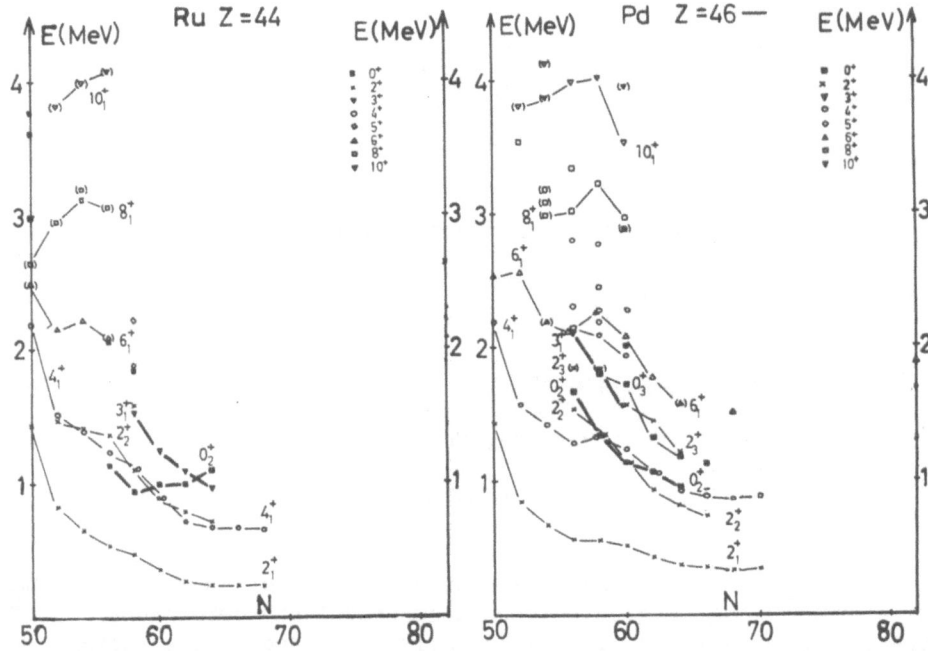

Fig. 1. Energy level systematics of positive parity states in Ru
and Pd isotopes.

states, only $E(4_1^+)/E(2_1^+)$ is shown, all have about the same value
between N=50 - N=60. (This also holds if Cd nuclei are incorporated,
while Mo nuclei show a minimum at N=56.) The deviation from the
harmonic vibrator, H.V., value towards the middle of the shell are
larger for Ru than for Pd and this deviation increases with the
spin but the rotational values are not reached.

When states away from the yrast states are considered more
dramatic changes occur. The values for $E(0_2^+)/E(2_1^+)$ in the case of
Pd have a minimum at N=60 where the values approach the H.V. value.
The same trend is present in Cd nuclei with minimum given at
N=64-66 while Ru nuclei approach the H.V. value at N=56-58 followed
by a well pronounced increase in this ratio. $E(2_2^+)/E(2_1^+)$ have
similar trends as $E(0_2^+)/E(2_1^+)$ for both Ru and Cd and also for Pd
if the 928 keV state in ^{112}Pd is a 0^+ state.

An underlying vibrational character also implies that members
of the 3-phonon multiplett, 0^+, 2^+, 3^+, 4^+, and 6^+, might be
identified. In the case of Ru isotopes few such candidates have
been located. In Pd nuclei, on the other hand, many states with
correct spin and within the expected energy region are present.
The ratios $E(3_1^+)/E(2_1^+)$ are shown in fig. 2 and we note that **these**

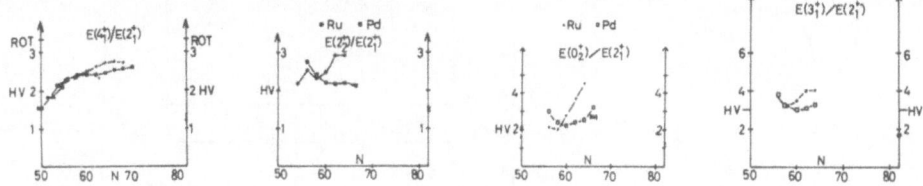

Fig. 2. Energy ratios for some states in Ru and Pd isotopes.

ratios have the same type of systematics as $E(2^+_2)/E(2^+_1)$. Unfortunately the information on the collective character of the states in the 3-phonon region is very scarce and furthermore it should be noted that these states are expected at about the energy of the energy gap and thus states with particle degrees of freedom as well as mixtures of such freedoms with collective ones are to be expected.

Information on high spin states in Ru and Pd nuclei is only available from (heavy ion, xn)-reactions. It is quite clear from the discussion above that results from heavy ion Coulomb excitation and thus information on higher states in the more neutron rich Ru and Pd nuclei would be most valuable. Grau and coworkers[2] in their discussion of (^{13}C,nγ) reactions leading to excited states in $^{102-106}$Pd interpret the 10^+ states in the yrast band of ^{104,106}Pd to be two-quasiparticle states and the higher states to come from the core coupled to this band-head. In the following high-spin states will not be discussed.

The interpretation of some of the nuclei in this mass region to have an underlying vibrational character, rather quickly broken in the case of Ru isotopes, is also in accordance with electro-magnetic properties. Absolute B(E2)-values are mainly available for the states 0^+_2, 2^+_2 and 4^+_1 and the experimental results are shown in fig. 3. Values of the quadrupole moments of the first excited 2^+ states, $Q(2^+_1)$, are given in fig. 4. The sign of the Coulomb interference term, $P_3 = M_{0_1 2_1} + M_{2_1 2_2} + M_{2_2 0_1}$, has been experimentally determined to be negative for ^{108,110}Pd[3] and ^{102,104}Ru[4,5]. This sign is in accordance with collective nuclear model predictions and only values of $Q(2^+_1)$ corresponding to $P_3 < 0$ are given.

An examination of fig. 4 shows that the values of $Q(2^+_1)$ are negative and generally about half the rotational values and that $Q(2^+_1)$ for ^{104}Ru is very close to that value. B(E2)-ratios for "allowed transitions" are normally smaller than the value given by the H.V. and B(E2) for "not allowed transitions" are just a few percent of the value of $B(E2; 2^+_1 \rightarrow 0^+_1)$. $B(E2; 0^+_2 \rightarrow 2^+_1)$ is almost constant and of the same magnitude as $B(E2; 2^+_1 \rightarrow 0^+_1)$. B(E2)-values to the 0^+_2 states should however be treated with some care since

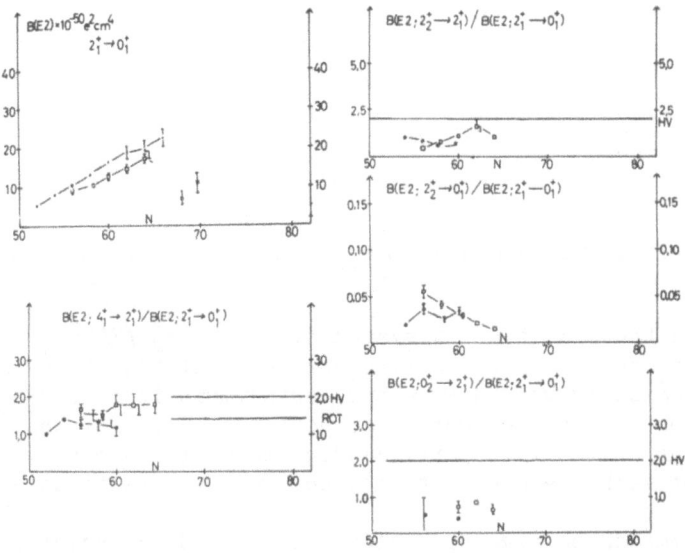

Fig. 3. Experimental B(E2)-values in Ru · and Pd□ . The results
         are taken from ref. 6 and ref. 5 for Ru isotopes and
         from ref. 7 and ref. 8 for Pd isotopes.

calculations show that realistic matrix elements within the 2-
phonon multiplett in the case of [104]Pd have an influence on the
Coulomb excitation cross-section to the 0_2^+ state with as much as
30 %. Generally the known B(E2)-ratios in Ru are somewhat smaller
than the corresponding ratio in Pd nuclei.

The energies of the three-phonon states can easily be
calculated to lowest order in the anharmonic vibrational limit[19].
For Pd nuclei such calculations give differences in the order of
300-400 keV to the experimental values of possible candidates,
except for 6^+ states. There are no absolute B(E2)-values known to
these states, but in a few cases mixing ratios have been measured
for transitions from the states. The percentage E2 in the transi-
tions $3_1^+ \rightarrow 2_1^+$ and $3_1^+ \rightarrow 2_2^+$ has been found to be 70 % or more[2,20-23]
and values of B(E2) to the n=2 states are larger than to n=1.
Taking this as an indication of collectivity the energy differences
given above indicate a three-phonon interaction of a large
magnitude[19].

Transitions within the two-phonon multiplet are strictly
forbidden in the H.V. limit. For an anharmonic vibrator, on the
other hand, such matrix elements should excist and can even be
expected to attain quite large values. Experimentally they are
difficult to measure. For [108]Pd Stelson and coworkers[24] tried to

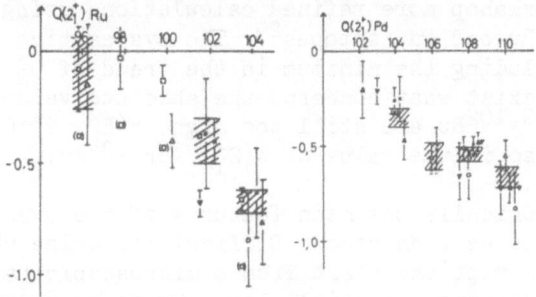

Fig. 4.   Experimental results for $Q(2_1^+)$ in Ru and Pd isotopes.
Meanvalues are given. The results are from: For Ru
isotopes o ref. 9 x ref. 20 ● ref. 11 ◻ ref. 12 ▲ ref. 5
▼ ref. 13. (◻) relative $Q(2_1^+)^{102}$Ru=-0.68. For Pd isotopes
o ref. 7 x ref. 14 ● ref. 15 ◻ ref. 16 ▲ ref. 17 ▼ ref 3.
▲ ref. 8 ▼ ref. 12. The value on ^{108}Pd from ref. 16 is
corrected following ref. 18.

detect the $4_1^+ \rightarrow 2_2^+$ γ-ray transition. The transition could not be
detected but they found an upper limit which corresponds to
$B(E2; 4_1^+ \rightarrow 2_1^+)/B(E2; 4_1^+ \rightarrow 2_2^+) > 85$, which in a simple treatment of
anharmonic effects taking only leading order terms into considera-
tion is one order of magnitude too large[19].

Most theoretical calculations on even Ru and Pd isotopes a few
neutrons away from the closed shell have been performed starting
from the underlying vibrational character[25-27]. Discrepancies
between theory and experiment do exist but the main features are
well reproduced.

Using the "old type" of the "Interacting Boson Approximation",
IBA, fits have been made to $^{100-104}$Ru[5] and $^{102-110}$Pd[8,28]. There
were seven parameters to be fitted which leave only a few direct
comparisons to be made. In the case of Pd isotopes the agreement
between theoretical and experimental quantities were in most cases
excellent and the parameters changed in a smooth way. For $^{100-104}$Ru
fits were made using as input values one- and two-phonon changing
terms from the trends obtained for the Pd isotopes. In this way it
was possible to obtain good fits for ^{100,102}Ru but not for ^{104}Ru,
which had to be fitted separately. Generally it was found easy to
obtain a good agreement with the yrast states but more difficult
in the case of 3_1^+ and 0^+ states. This is in agreement with the
conclusions made by Lie and Holtzwarth in their calculations on
^{100,102}Ru[27]. They conclude that the yrast states are quite
insensitive to the details of the theory but that 0^+ states, on
the other hand, are sensitive. They also suggest that the quadrupole phonon approach can not explain the low lying 0^+ states.

At this workshop more refined calculations using IBA are presented for Ru and Pd isotopes[29]. The systematics are well reproduced including the minimum in the trend of 0_2^+ in Ru isotopes. Discrepancies exist what concerns the absolute values. $E(0_2^+)$ in the case of ^{100,102}Ru are still too high, $B(E2; 2_2^+ \to 0_1^+)$ is under-estimated and so is the value of $Q(2_1^+)$ for ^{104}Ru.

Phenomenologically the main features of the even Ru and Pd isotopes seem to be understood. Difficulties arise when the theoretical descriptions start from a microscopic basis and the positions of the single-particle levels must be known with high accuracy[25,27]. Inclusion of particle excitation modes might be important[25] and will further complicate the calculations. Obviously more experimental data especially on neutron-rich Ru isotopes are needed in order to make a detailed comparison with theoretical predictions possible.

REFERENCES

1. N. Kaffrel, Proc. 3rd Int. Conf. on nuclei far from stability, Cargèse, 1976, 483
2. A. Graue, L.E. Samuelson, F.A. Rickey, P.C. Simms and G.J. Smith, Phys. Rev. C14 (1976) 2297
3. L. Hasselgren, C. Fahlander, L.O. Edvardson, J.E. Thun, B.S. Ghuman and B. Skaali, Nucl. Phys. 264 (1976) 341
4. C. Fahlander, L. Hasselgren, J.E. Thun, A. Bockisch, A.M. Kleinfeld, A. Gelberg and K.P. Lieb, Phys. Lett. 60B (1976) 342
5. C. Fahlander, L. Hasselgren, G. Possnert and J.E. Thun, Physica Scripta 18 (1978) 47
6. F.K. Mc Gowan, R.L. Robinson, P.H. Stelson and W.T. Milner, Nucl. Phys. 113 (1968) 529
7. R.C. Robinson, F.K. Mc Gowan, P.H. Stelson, W.T. Milner and R.O. Sayer, Nucl. Phys. 124 (1969) 553
8. L. Hasselgren, C. Fahlander, J.E. Thun, B. Orre and N.G. Jonsson, UUIP-957 (1977)
9. M.F. Nolan, I. Hall, D.J. Thomas and M.J. Throop, J. Phys A6 (1973) 57
10. A.M. Kleinfeld, private communication
11. P.H. Stelson, priv. comm. to J. de Boer and J. Eichler in Adv. in Nucl. Phys. 1 (1968) 1
12. M. Magnard, D.C. Palmer, J.R. Cresswell, P.D. Forsyth, I. Hall and D.G. Martin, J. Phys. G3 (1977) 1735
13. A. Bockisch, private communication
14. A. Christy, I. Hall, R.P. Harper, I.M. Naqib and B. Wakefield, Nucl. Phys A142 (1970) 591
15. D. Ward, J.S. Geiger and R.L. Graham, Bull. Am. Phys. Soc. 16 (1971) 14
16. R.P. Harper, A. Christy, I. Hall, I.M. Naqib and B. Wakefield, Nucl. Phys. 162 (1971) 161
17. W.R. Lutz, J.A. Thomson, R.P. Scharenberg and R.D. Larsen, Phys. Rev. C6 (1972) 1385

18. I.M. Naqib, A. Christy, M.F. Nolan and D.J. Thomas, J. Phys. G3 (1977) 507
19. A. Bohr and B.R. Mottelson, in Nuclear Structure, Vol. II, Benjamin Inc., 1975
20. J. Lange, A.T. Kandil, J. Neuber, C.D. Uhlhorn, H. von Buttlar and A. Bockisch, Nucl. Phys. 292 (1977) 301
21. B. Singh and H.W. Taylor, Nucl. Phys. 155 (1970) 70
22. P.J. Tivin, B. Singh and H.W. Taylor, J. Phys. G3 (1977) 1267
23. K. Sümmerer, N. Kaffrell and N. Trautmann, Nucl. Phys. 308 (1978) 1
24. P.H. Stelson, S. Raman, J.A. McNabb, R.W. Lide and C.R. Bingham, Phys. Rev. C8 (1973) 368
25. T. Kishimoto and T. Tamura, Nucl. Phys. A270 (1976) 317 and T. Kishimoto, T. Tamura and K. Weeks, private communication
26. D. Janssen, Proc. Topical Conf. on problems of vibrational nuclei, Zagreb 1974, ed. G. Alaga, V. Paar and L. Sips (North-Holland, Amsterdam, 1975)
27. S.G. Lie and G. Holtzwarth, Phys. Rev. C12 (1975) 1035
28. O. Scholten, private communication
29. O. Scholten, contribution at this workshop

B(E2)-VALUES OF HIGH SPIN STATES IN ^{158}DY

Hans Emling

Gesellschaft für Schwerionenforschung GSI

Postfach 541, 6100 Darmstadt 1, West Germany

One of the main aspects of the interacting boson approximation is the explicitly included finite number of nucleons. This
finite number reduces the collective behavior of a nucleus and
becomes a dominant feature at high spins. A definite test of this
prediction is the behavior of the B(E2)-values in the ground state
band of rotational like nuclei. The SU(3)-approximation predicts
in zero order a cutoff-factor (2N-J) (2N+J+3) (J is the nuclear
spin and 2N the number of nucleons outside the closed shell).

In the following, the results of a B(E2)-measurement of high
spin states in ^{158}Dy are presented[1]. In this nucleus the high
spin states are excited by a compound nucleus reaction
^{26}Mg(^{136}Xe,4n) and the large recoil velocity of v/c = 0.08
allowed to measure the lifetimes of all these states by a Plunger-
recoil distance device although the lifetimes for the highest
spins are in the order of 1 ps. The highest observed spin is
the J = 26$^+$.

In fig. 1 the normalized ratio B(E2)$_{exp}$/B(E2)$_{rot}$ is shown.
The E2-transition probabilities calculated from the interacting
boson model[2] in the SU(3) limit drop down to zero for the transition J = 28$^+$→J = 26$^+$ assuming that the acting number of nucleons
in^{158} Dy is 2N = 26. The experimental values seem to be reduced
for the upper spins, but this reduction is overestimated in the
interacting boson model. The enhancement of the low spin B(E2)-
values (J = 4$^+$, 6$^+$, 8$^+$), however, indicates that centrifugal
stretching occurs. Therefore higher order perturbation terms have
to be applied[2].

The reduction of the high spin B(E2)-values, however, not nec-

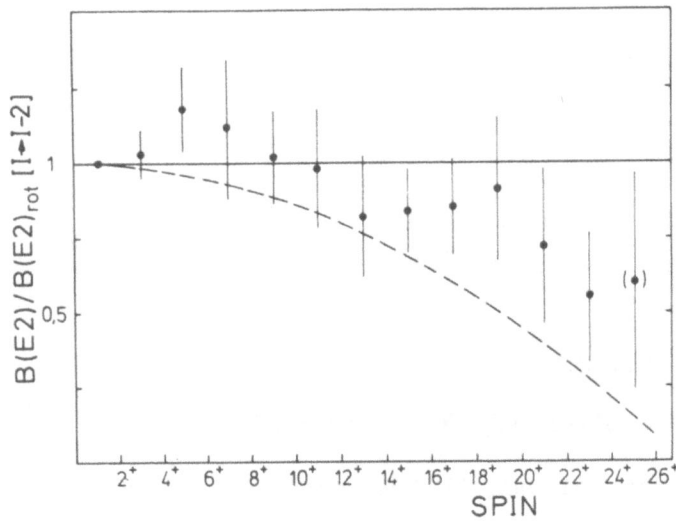

Fig. 1. Normalized B(E2)-values in ^{158}Dy.
Dashed curve corresponds to the interacting boson
approximation in the rotational limit.

essarily has to be interpreted as an effect due to the finite
nucleon number as predicted by the interacting boson model. The
^{158}Dy energy level spacing shows a small deviation from the J(J+1)
rotational rule. This upbending effect was described by Sano
et al.[3] as a change from the superfluid to the normal phase. In
this model the vanishing neutron pair energy gap is accompanied
with a B(E2)-value reduction around spin J = 14$^+$ and the vanishing
proton energy gap is responsible for the reduction above spin
J = 20$^+$. More likely, however, the upbending is explained as the
rotational alignment of an $i_{13/2}$ neutron pair, the experimental
B(E2)-values then require that the intrinsic quadrupole moment of
the aligned band is changed by 15% compared to the ground state
band.

    Probably many effects of different physical nature can pro-
duce a reduction of the electromagnetic transition probabilities
at high spin and only systematic experimental investigations will
lead to unique results. Such experiments are presently carried
out at GSI using heavy ion beams in inverse HI,xn reactions as
described above or alternately multiple Coulomb excitation.

References

1. H. Emling, P. Fuchs, E. Grosse, D. Husar, D. Schwalm, R.S. Simon, H.J. Wollersheim (GSI Darmstadt) and D. Pelte (University Heidelberg), to be published.
2. A. Arima and F. Iachello, Ann. Phys. 111 (1978), 201.
3. M. Sano, T. Takemasa and M. Wakai, J. Phys. Soc. Japan, Suppl. 34 (1973), 365.

SHELL MODEL DESCRIPTION OF INTERACTING BOSONS

Igal Talmi

The Weizmann Institute of Science

Rehovot, Israel

The interacting boson model[1,2] gives a very good and unified description of collective states in medium and heavy nuclei. The Hamiltonian of this model conserves the number of bosons which suggests that they represent in some way (valence) nucleon pairs. More specifically, proton pairs and neutron pairs should be considered. A more detailed model involving proton and neutron s- and d-bosons was introduced[3] and the interacting bosons were considered in the framework of the shell model. The relationship between these models given in ref. 3 will be discussed in the present talk.

As well known, the shell model is a very good approximation to the complex nuclear many-body problem. In light nuclei and in heavier nuclei near closed shells, it is possible to use it for successful calculations of energy levels and transition probabilities. This is accomplished by using effective two-body interactions and effective single particle operators for electro-magnetic transitions. These quantities have been determined from experiment in those cases where the data satisfied consistency checks imposed by the model[4].

It is clear, however, that in cases with many valence nucleons, the straight-forward application of the shell model becomes prohibitively difficult. As the number of active orbits - those that participate in the shell model configurations - grows, the number of matrix elements of the effective interaction increases very rapidly. They cannot be determined by the experimental data, and this does not enable us to write down the sub-matrices of the shell model Hamiltonian which should be diagonalized. These matrices be-

come so large that it becomes impossible to diagonalize them with
present day computers. More important, those matrices contain
information about very many states which lie at high energies
where excited configurations are certainly important for their
description. We are mainly interested in low-lying levels which
show very simple and regular features. A computer print-out of
a wave function, containing thousands and thousands of components,
gives us no physical insight about the emergence of these features.

Consider for example the case of $^{112}_{50}\mathrm{Sn}_{62}$ where the 12 valence
neutrons occupy the $1g_{7/2}$, $2d_{5/2}$, $2d_{3/2}$, $3s_{1/2}$ and $1h_{11/2}$ orbits.
The shell model configurations arising from 12 neutrons in these
orbits give rise to

$$56,907 \quad \text{states with J=0}$$

$$267,720 \quad \text{states with J=2}$$

$$\text{and } 426,558 \quad \text{states with J=4.}$$

Still, up to excitation energy of 2.25 MeV there are only 4 states
(Fig. 1).

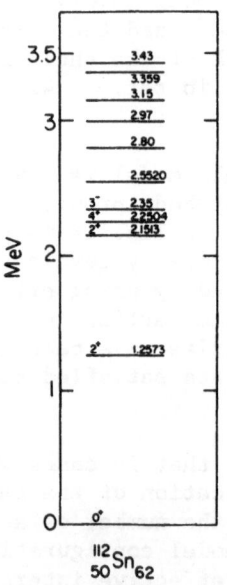

Fig. 1. Low lying levels of $^{112}_{50}\mathrm{Sn}$.

Some of these states have very simple properties, and it is in these states that we are interested. For example, binding energies of Sn isotopes obey a simple mass formula and the 0-2 spacing is fairly independent of the neutron number. In nuclei with both valence protons and valence neutrons there is a tremendous increase in complexity. In a typical nucleus like $^{154}_{62}Sm_{92}$ the 12 protons occupy the active orbits listed above for the valence neutrons in Sn isotopes. The active orbits for the 10 valence neutrons are $1h_{9/2}$, $2f_{7/2}$, $2f_{5/2}$, $3p_{3/2}$, $3p_{1/2}$ and $1i_{13/2}$. In the shell model, in configurations arising from these active orbits there are

$$41 \ , \ 654 \ , \ 193 \ , \ 516 \ , \ 797 \quad \text{states with J=0}$$

$$346 \ , \ 132 \ , \ 052 \ , \ 934 \ , \ 889 \quad \text{states with J=2}$$

$$\text{and} \quad 530 \ , \ 897 \ , \ 397 \ , \ 260 \ , \ 575 \quad \text{states with J=4} \ .$$

It is obvious that a complete shell model calculation is beyond any hope. We must introduce some truncation scheme or in better words, a coupling scheme which will give a simplified yet adequate description of the low-lying levels. These levels usually have some collective properties and are grouped into various vibrational, transitional or rotational bands.

For the case of semi-magic nuclei, like the Sn isotopes, good description of low-lying levels was given in terms of generalized seniority[5] . The operator

$$s^+_j = \frac{1}{2} \sum_m (-1)^{j-m} a^+_{jm} a^+_{j-m} \qquad (1)$$

creates a j^2 J=0 pair. In the case of several j-orbits, the operator

$$s^+ = \sum_j \alpha_j s^+_j \qquad (2)$$

creates a correlated J=0 pair. Consider a shell model Hamiltonian for which $(s^+)^n |0\rangle$ are eigenstates, namely

$$H(s^+)^n |0\rangle = E_n (s^+)^n |0\rangle \qquad (3)$$

(here $|0\rangle$ is the state with closed shells). If (3) holds for n=1 and n=2, we obtain the conditions

$$Hs^+ |0\rangle = V_o \ s^+ |0\rangle \qquad (4)$$

$$\left[\left[H, s^+ \right] , s^+ \right] = \Delta (s^+)^2 \qquad (5)$$

where V_o and Δ are constants constructed of matrix elements of H. It then follows that for any n we have

$$H(S^+)^n \,|0\rangle = (nV_o + \frac{n(n-1)}{2}\,\Delta)\,(S^+)^n\,|0\rangle \qquad (6)$$

Thus, binding energies are given by a linear and quadratic function of n with constant coefficients in agreement with experiment. The configuration mixings introduced this way involve all active orbits and no sub-shell effects show up (e.g. Fig. 2). These states are defined to have generalized seniority v=0.

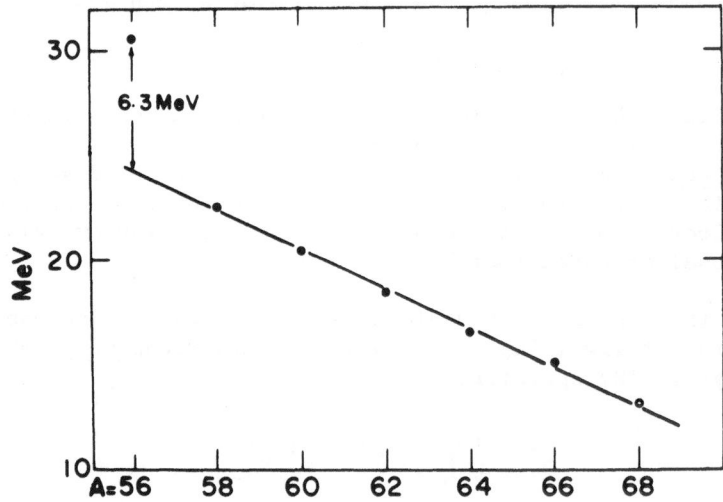

Fig. 2. Neutron pair separation energies in Ni isotopes.

The same eigenvalues can be obtained from a Hamiltonian constructed by using <u>boson</u> creation and annihiliation operators

$$V_o s^+ s + \frac{1}{2}\,\Delta (s^+)^2 (s)^2 \qquad (7)$$

for the set of states $(s^+)^n\,|0\rangle$. This is certainly not a "boson expansion" of the shell model Hamiltonian. The boson operators are not equal to fermion pair operators, nor does the boson Hamiltonian contain 4 fermion interactions. It is a model of the shell model Hamiltonian for certain states. Thus, we establish a correspondence between certain <u>states</u> and we do not have an expansion of operators which must be valid for all millions of states.

For the J=2 levels we consider the operators

$$D^+_{jj',\mu} = \sum_{mm'} (jmj'm'|jj'2\mu) \, a^+_{jm} a^+_{j'm'} \tag{8}$$

which create jj' J=2 pair states. The operator

$$D^+_\mu = \sum_{jj'} \beta_{jj'} D^+_{jj',\mu} \tag{9}$$

creates a <u>correlated</u> J=2 pair. If now $D^+(S^+)^{n-1}|0\rangle$ are eigenstates of the shell model Hamiltonian, the following additional conditions must be satisfied

$$HD^+_\mu |0\rangle = V_2 D^+_\mu |0\rangle \tag{10}$$

$$\left[\left[H,S^+\right],D^+_\mu\right] = \Delta S^+ D^+_\mu \tag{11}$$

It then follows that for any n

$$HD^+_\mu(S^+)^{n-1}|0\rangle = (nV_0 + \frac{n(n-1)}{2}\Delta + V_2 - V_0)D^+_\mu(S^+)^{n-1}|0\rangle \tag{12}$$

Thus, 0-2 spacings are independent of n in agreement with experiment (Fig. 3 and Fig. 4). It can also be shown that these special J=2 states exhaust the E2 sum rule in the shell model space considered.

Fig. 3. Systematics of $J=2^+$ levels in Sn isotopes.

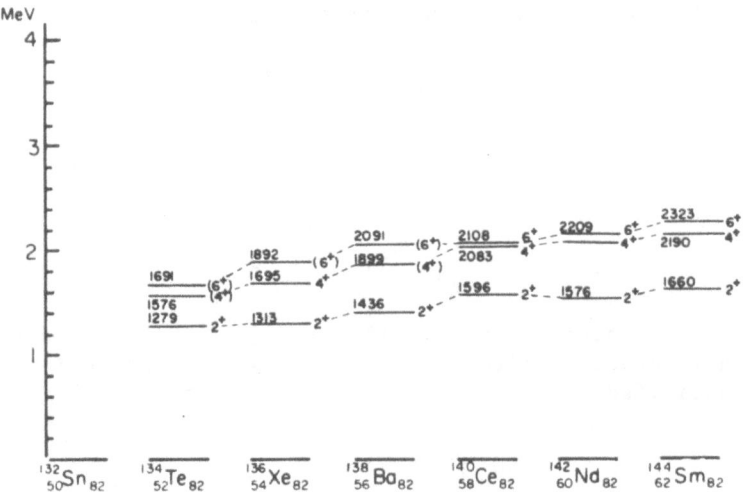

Fig. 4.   Systematics of J-2$^+$ levels in isotones with N=82.

These states, with generalized seniority v=2, $D_\mu^+(S^+)^{n-1} |0\rangle$
correspond to boson states $d_\mu^+(s^+)^{n-1} |0\rangle$. The correspondence is
somewhat more complicated for higher numbers of d-bosons. States
like $(D^+ \times D^+)^{(J)} (S^+)^{n-2} |0\rangle$ contain lower seniority components
which must be projected out. The resulting states with generalized
seniority v correspond to boson states $(d^+)^{v/2} (s^+)^{n-v/2} |0\rangle$.

These statements can be made more precise for a single j-shell
or for the case where all α_j in (2) are equal. In that case, generalized seniority is replaced by seniority[6] and the latter supplies
a complete characterization of states. The non-vanishing compon-
ents with highest seniority in the nucleon states $(D^+)_{\gamma JM}^{v/2} (s^+)^{n-v/2} |0\rangle$
have seniority v and are eigenstates of the pairing interaction. Any
additional quantum numbers necessary to uniquely define the states
are denoted by γ. The corresponding boson states are $(d^+)_{\gamma JM}^{n_d}(s^+)^{n_s} |0\rangle$
where $v = 2n_d$ and $n - 2n_d + 2n_s$. All these states should be nor-
malized but for brevity we omit here the normalization factors. The
boson operator which is equivalent to the pairing interaction is
simply

$$2\Omega n_s - 2n_s(n_s-1) - 4n_s n_d \qquad\qquad (13)$$

where $2\Omega = \Sigma(2j+1)$, is the number of nucleon states in the shell
considered.

In semi-magic nuclei a simple coupling scheme for low-lying
levels is thus offered by generalized seniority. This does not
mean of course, that the simple minded pairing interaction is a
good approximation of the effective interaction. There are many

shell model Hamiltonians, which are much more general, that give rise to eigenstates with good generalized seniorities. This conservation of seniority is drastically changed when there are both protons and neutrons outside closed shells. The proton–neutron interaction certainly cannot preserve seniority. It is evidenced for instance, by the considerable reduction of the 0–2 spacings in such nuclei. This can be demonstrated in a simple case involving $1g_{9/2}$ configurations with the same seniority v have spacings independent of n as seen in nuclei with neutron number 50 (Fig. 5). Once there are valence protons and neutrons this is dramatically changed (Fig. 6). Indeed, the proton–neutron interaction derived from such nuclei leads to considerable change in the 0–2 separation for calculated T=0 states in the $g_{9/2}^4$ configuration as compared to the T=2 states (Fig. 7). The spacings of the latter are the same as in the $g_{9/2}^2$ T=1 states. An interaction which manifestly breaks seniority must contain scalar products of even tensors. An acceptable interaction of this kind is the quadrupole–quadrupole interaction.

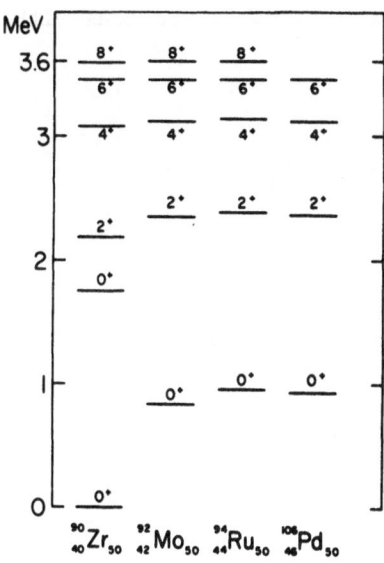

Fig. 5. Levels of even $g_{9/2}^n$ configurations in N=50 nuclei.

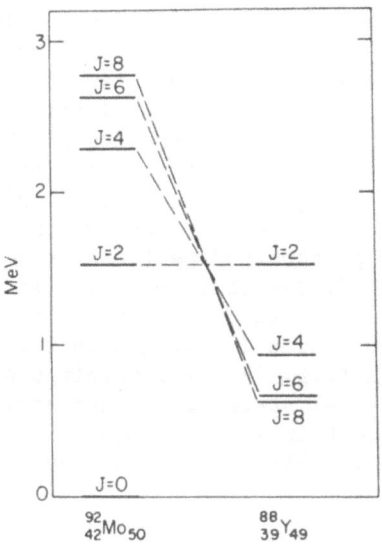

Fig. 6.   Levels of some $g_{9/2}^n$ configurations with protons and
          neutrons.

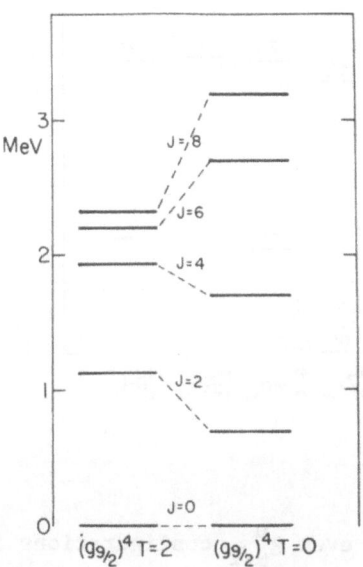

Fig. 7.   Calculated levels with T=0 and T=2 of the $g_{9/2}^4$
          configuration.

"Pairing plus P_2 interaction" has been a popular slogan for many years. It makes sense, however, only if pairing or rather seniority, determines proton and neutron states whereas the P_2 force acts only between protons and neutrons. The proton-neutron interaction is strong and attractive and leads to the well-known collective features of nuclei.

States of nuclei with both valence protons and neutrons can be simply constructed from states of the valence protons coupled to states of the valence neutrons. We have considered above a set of states for the protons (or the neutrons) for which a correspondence can be established with boson states. It turns out that the states of the proton neutron system strongly admixed by the quadrupole-quadrupole interaction are precisely those constructed from the boson-like states of the protons coupled to boson-like states of the neutrons. The model we propose is to truncate the shell model Hamiltonian, including the proton-neutron interaction, into the submatrix constructed by using only these sets of states. Furthermore, in order to avoid the complexity of the fermion operators, we shall use boson states and equivalent Hamiltonian and quadrupole operator.

We have already considered those parts of the Hamiltonian involving separately protons and neutrons. Our main attention should be focused on the proton-neutron interaction. We have to demonstrate that the truncation we propose is indeed a good approximation. In the next step we have to show that matrix elements of V_{pn} between nucleon states are well approximated by corresponding matrix elements between boson states.

Matrix elements between nucleon states are given by[6]

$$< \alpha_p J_p \alpha_n J_n JM | T_p^{(2)} \cdot T_n^{(2)} | \alpha_p' J_p' \alpha_n' J_n' JM > =$$

(14)

$$= (-1)^{J_n + J_p' + J} \begin{Bmatrix} J_p & J_n & J \\ J_n' & J_p' & 2 \end{Bmatrix} (\alpha_p J_p | T_p^{(2)} | \alpha_p' J_p') (\alpha_n J_n | T_n^{(2)} | \alpha_n' J_n')$$

Thus, we have to consider matrix elements of the quadrupole operator for the proton configuration and for the neutron configuration. The general situation, with several j-orbits having different α_j values, is very complex. Straight-forward calculations can be carried out for a single j-orbit. Examples of such calculations will be shown by Otsuka[7]. It seems that in the case of a large j-orbit matrix elements of this V_{pn} interaction leading out of the model space are rather small. Of particular interest is the matrix element connecting the states with J=2, v=2 (one d-boson state) and the state with J=4, v=2 which is <u>outside</u> the model space.

The dependence of this matrix element on the total number of (identical) nucleons, as well known, is given by

$$\frac{2j+1-2n}{2j+1-2v} \ (j^2 \ J=2 \ | \ \|T^{(2)}\| \ | \ j^2 \ J=4) \qquad (15)$$

for v=2. Thus, as we add more nucleons this matrix element goes down and vanishes in the middle of the shell.

In this simple example (a single j-orbit), it is possible to see how well we can replace the complicated calculations in the nucleon space by using equivalent boson states and operators. As an example, we compare the ratio between

$$(j^4 \ v=4 \ J \ | \ \|T^{(2)}\| \ | \ j^4 \ v=2 \ J=2)/(j^2 \ J=2 \ | \ \|T^{(2)}\| \ | \ j^2 \ J=0) \qquad (16)$$

and the corresponding ratio for bosons (where, for simplicity, indices of components of operators as well as normalization of states were omitted)

$$<0 \ |(d \times d)^{(J)} \ (d^+ s + s^+ d) d^+ s^+ \ |0> / <0 \ |d \ (d^+ s + s^+ d) \ s^+ \ |0> \qquad (17)$$

This ratio is given by the following expression

$$\left[1 + \frac{4}{2j-3} - \frac{50(2j+1)^2}{(2j-3)^2} \begin{Bmatrix} j & j & 2 \\ j & j & 2 \\ 2 & 2 & 2 \end{Bmatrix} - \frac{200(2j+1)^2}{(2j-3)^3} \begin{Bmatrix} j & j & 2 \\ 2 & 2 & j \end{Bmatrix}^2 \right]^{1/2} \qquad (18)$$

The terms following 1 in the brackets arise from the commutation re-lations between the fermion operators in D and S. The last term is due to the necessity of projecting out lower seniority components. The other terms are due to the Pauli principle. For a realistic case, the value of j should be large so that the Pauli principle will not be too restrictive. It turns out that for J=2 (and very similarly for J=4) the corrections become small even for j=13/2. For j=13/2 up to j=31/2 the corrections do not amount to more than 1%. For J=0 the Pauli principle is much more important and even for j=31/2 the corrections are still about 10%.

It is clear, however, that the Pauli principle will become very important as the number of nucleons increase. Its effects cannot be simply reproduced by using single boson operators and boson states. For example, the dependence on n of (15) is not obtained by using the single boson operators $(d^+ \times d)^{(2)}$. In fact, matrix elements of the latter remain unchanged as the number of s-bosons is increased. As a result, we have to introduce this dependence explicitly into

the equivalent boson operators. In the case of a single j-orbit, or
equal α_j coefficients in (2), this is a simple matter. When calculating matrix elements of the quadrupole operator between nucleon
states with the same seniority v, the equivalent boson operator
should not be simply $(d^+ \times d)^{(2)}$. The equivalent boson operator
should be

$$\frac{2\Omega - 4n_s - 4n_d}{2\Omega - 4n_d} (d^+ \times d)^{(2)} \tag{19}$$

where $n_s = s^+ s$ and $n_d = \Sigma d_\mu^+ d_\mu$ are the s- and d-boson number operators, respectively.

The behaviour of matrix element connecting states with different seniorities v and v-2 is quite different. Between states with
v=4 (two di-bosons) and v-2=2 (one d-boson) it is given by

$$\sqrt{\frac{(n-v)(2j+3-n-v)}{2(2j+3-2v)}} \; (j^4 \; v=4 \; J \,|\, \|T^{(2)}\| \,|\, j^4 \; v=2 \; J=2) \tag{20}$$

for v=4. The $\sqrt{n-v}$ behaviour in (20) is obtained also from a single
boson operator and boson states. The $\sqrt{2j+3-n-v}$ factor, however, is
due to the Pauli principle and therefore must be explicitly introduced into the boson operators. For calculating matrix elements
between states with different seniorities the equivalent quadrupole
operator should not be single boson operator $d^+ s + s^+ d$. It should
instead be given by

$$\sqrt{\frac{2\Omega + 2 - 2n_s - 4n_d}{2\Omega + 2 - 4n_d}} \; d^+ s + s^+ d \; \sqrt{\frac{2\Omega + 2 - 2n_s - 4n_d}{2\Omega + 2 - 4n_d}} \tag{21}$$

Thus, the quadrupole operator for the protons (or the neutrons)
which appears in the proton-neutron interaction should be replaced
in the equivalent boson Hamiltonian by

$$\alpha \frac{2\Omega - 4n_s - 4n_d}{2\Omega - 4n_d} (d^+ \times d)_\mu^{(2)} + \tag{22}$$

$$+ \beta(\sqrt{\frac{2\Omega + 2 - 2n_s - 4n_d}{2\Omega + 2 - 4n_d}} \; d_\mu^+ s + s^+ d_\mu \sqrt{\frac{2\Omega + 2 - 2n_d - 4n_d}{2\Omega + 2 - 4n_d}})$$

where n_s and n_d are the number operators for s- and d-bosons respectively. The coefficients α and β can be simply calculated
for a single j-orbit. In general, they can be treated as parameters. From the point of view of boson models, the resulting

Hamiltonian seems a very complicated one. On the other hand, if
viewed as an approximation of the shell model Hamiltonian, its
form appears as a natural consequence.

The model consists now of four kinds of bosons. Proton and
neutron s- and d-bosons. The boson Hamiltonian should be diagonalized in the space of states obtained by coupling a definite
number of proton bosons (the number of valence protons divided by
two) and a similarly defined number of neutron bosons. This
scheme may look complicated, but it should always be compared to
the shell model situation with billions of states. Another important point is that in limiting cases, the picture may become even
simpler. In particular, the IBA bosons considered by Arima and
Iachello[2] can be easily constructed in the present model. They
correspond to couplings of proton and neutron bosons which are
fully symmetric. The states $(s_p^+)^m (s_n^+)^n \,|0\rangle$ correspond to
$(s^+)^{m+n} \,|0\rangle$ of IBA. The states

$$(d_p^+ s_p + d_n^+ s_n) \; (s_p^+)^m (s_n^+)^n \,|0\rangle \tag{23}$$

correspond to the $d^+ (s^+)^{m+n-1} \,|0\rangle$ states of IBA. In general, the
states

$$(d_p^+ s_p + d_n^+ s_n)_{\gamma JM}^{n_d} \; (s_p^+)^m (s_n^+)^n \,|0\rangle \tag{24}$$

correspond to the

$$(d^+)_{\gamma JM}^{n_d} (s^+)^{n_s} \,|0\rangle \qquad\qquad n_d + n_s = m + n \tag{25}$$

states of IBA. We have introduced a special quantum number, F-spin,
to describe this symmetry[3] . It should be remembered, however,
that F-spin symmetry may be a good approximation only for certain
nuclei. In general, both proton bosons and neutron bosons must be
explicitly considered.

Preliminary results using this approach have been carried out
by Otsuka[7] and yield qualitative agreement with the data. Very
good agreement with the spectra of Ba isotopes can be obtained by
slight changes of the parameters. Such renormalizations could
arise from the effect of states outside the model space. The calculations of Scholten[8] also reproduce for other cases, like the
Sm isotopes, the results of IBA which agree well with experiment.
In fitting the data of various isotopes, the parameters change
roughly in accordance with (22).

The most attractive feature of the interacting boson model is
its ability to describe collective states in nuclei in both the

vibrational and rotational regions, as well as the transition between them. It seems now that the use of interacting bosons may
be a good approximation to complex shell model calculations. We
therefore believe that we have at our hands a shell model description of collective states in nuclei which is realistic enough to
contain the important ingredients of many-orbits shell model calculations and, at the same time, sufficiently simple to allow
actual successful calculations. The model is also simple enough
to allow our physical intuition to grasp the main features of collective states and the way in which they arise in the nuclear
shell model.

REFERENCES

1. F. Iachello in Proc. 1974 Amsterdam Conf. on Nucl. Struct.
   and Spectroscopy (Scholar's Press, Amsterdam 1974) p. 163.
   F. Iachello and A. Arima, Phys. Lett. 53B (1974) 309.
2. A. Arima and F. Iachello, Phys. Rev. Lett. 35 (1975) 1069;
   Ann. of Phys. 99 (1976) 253, and Ann. of Phys. 111 (1978) 201.
3. A. Arima, T. Otsuka, F. Iachello and I. Talmi, Phys. Lett. 66B
   (1977) 205 and Phys. Lett. 76B (1978) 139.
4. A summary of this approach is given by I. Talmi, Rev. Mod.
   Phys. 34 (1962) 704.
5. I. Talmi, Nucl. Phys. A178 (1971) 1, S. Shlomo and I. Talmi,
   Nucl. Phys. A198 (1972) 81, I. Talmi, Rivista Nuovo Cimento 3
   (1973) 85.
6. See e.g. A. de-Shalit and I. Talmi, Nuclear Shell Theory,
   Academic Press, N.Y. (1963).
7. T. Otsuka, following lecture.
8. O. Scholten, in these proceedings.

Nuclear Shell Model and Interacting Bosons

Takaharu Otsuka

Department of Physics, Faculty of Science
University of Tokyo
Hongo, Tokyo, 113

I. Introduction

In this workshop we have seen that the interacting boson model provides us with a unified phenomenological description of vibrational, rotational and transitional nuclei.

The latest and most successful version of the interacting boson model is the proton-neutron boson model in which we have the proton s_π- and d_π-bosons, and the neutron s_ν- and d_ν-bosons. The success of this model in the phenomenological analysis suggests that these bosons, s_π, d_π, s_ν and d_ν, represent some underlying (approximated) symmetry in the fermion space. Actually this proton-neutron boson model was proposed from microscopic investigations of the s-d bosons. We are going to discuss in this talk a relation between the boson model and the microscopic nuclear structure.

Works presented in this talk were performed in collaboration with Arima, Iachello and Talmi, and were reported in refs. 1 ~ 4. A detailed account is given in ref. 4.

II. Boson approximation in a single j-orbit

We construct a boson system as an approximation to the shell model. The exact shell model calculation itself is not a useful tool to investigate medium-heavy nuclei except those near closed shells. This is because the dimension of the configuration space becomes extremely large, for example $\sim 10^{14}$ dimension in ^{154}Sm with the major shells 50-82 for protons and 82-126 for neutrons, as reported by Talmi in this workshop. Exact shell model calculations are then neither practical nor transparent. We first truncate the full shell model space to a small subspace, denoted S-D, and construct a boson system to describe the subspace in a good approximation. This approximation method enables us to perform sensible calculations as well as to offer physical insight into the nature of the low-lying collective states of even-even nuclei.

We first restrict ourselves to cases that protons and neutrons are occupying (large) single j-orbits respectively. Although the single j-orbit model is very schematic, one can enjoy an advantage to investigate relation between fermions and bosons without having large ambiguity of dynamical factors.

§2.1. Quasi-spin formalism for a single j-orbit

We begin by considering a system of n-even identical nucleons in a single j-orbit. We introduce creation $a^\dagger_{j,m}$ and modified annihilation operators $\tilde{a}_{j,m}=(-)^{j-m}a_{j,-m}$ in the usual way. From these we form pair creation, pair annihilation and one-body multipole operators as follows,

$$A^{\dagger(J)}_M = \frac{1}{\sqrt{2}}[a^\dagger_j a^\dagger_j]^{(J)}_M, \tag{2.1}$$

$$\tilde{A}^{(J)}_M = -\frac{1}{\sqrt{2}}[\tilde{a}_j \tilde{a}_j]^{(J)}_M, \tag{2.2}$$

$$U^{(J)}_M = [a^\dagger_j \tilde{a}_j]^{(J)}_M. \tag{2.3}$$

The nucleon number operator is

$$N^F = \sum_m a^\dagger_{j,m} a_{j,m} \tag{2.4}$$

and the degeneracy of the shell is $\Omega=j+\frac{1}{2}$. The three operators

$$S_+ = \sqrt{\Omega}\, A^{\dagger(0)}, \tag{2.5}$$

$$S_- = \sqrt{\Omega}\, \tilde{A}^{(0)}, \tag{2.6}$$

$$S_0 = \frac{1}{2}(N^F-\Omega) \tag{2.7}$$

satisfy the same commutation relations of the three components of an angular momentum vector [SU(2) algebra][5]. The vector $\vec{S}$ is called quasi-spin. The quasi-spin operators can be used to construct states with n particles and seniority v starting from states of v particles and seniority v,

$$|j^n,v,L,M> = \sqrt{\frac{(\Omega-\frac{1}{2}(n+v))!}{(\Omega-v)!(\frac{1}{2}(n-v))!}}\; S_+^{\frac{1}{2}(n-v)} \;|j^v,v,L,M>. \qquad (2.8)$$

The states $|j^v,v,L,M>$ satisfy

$$S_- \;|j^v,v,L,M> = 0. \qquad\qquad (2.9)$$

§2.2. Collective state vectors

We construct in the following the fermion (collective) sub-space S-D by using operators $S_+ (=\sqrt{\Omega}A^{\dagger(0)})$ in eq. (2.5) and $A^{\dagger(2)}$ in eq. (2.1).   We truncate the full shell model space to this S-D subspace (Fig. 1).   There is a problem of seniority projection in the space construction ;   Given a highest seniority state $|j^v,v,J,M>$ (possible additional quantum numbers are omitted), one can generate a state with v+2 particles by acting with $A_m^{\dagger(2)}$ on it:

$$\Psi\,(j^{v+2},L,M) = \mathcal{N}^{-1}\sum_m A_m^{\dagger(2)} \;|j^v,v,J,M-m>(J,2,M-m,m|L,M), \quad (2.10)$$

where $\mathcal{N}$ is the normalization constant.   However, this state involves generally components with seniorities v-2 and v in addition to v+2.   In order to remove the lower seniority components one can introduce an operator P:

$$P = \frac{1}{4S_0-6}(4S_0-6+S_+S_-)\;\frac{1}{2S_0-2}\;(2S_0-2+S_+S_-). \qquad (2.11)$$

The operator $D_m^{\dagger}$,

$$D_m^{\dagger} = P\,A_m^{\dagger(2)}, \qquad\qquad\qquad (2.12)$$

generate then only highest seniority states if they act on a highest seniority state.   Using the operators $D_m^{\dagger}$ we can construct a family of states

$$|j^v(D^{v/2}),\boldsymbol{\alpha},L,M> = \mathcal{N}_{v\alpha}^{-1}[[\ldots[D^{\dagger}D^{\dagger}]^{(L_1)}D^{\dagger}]^{(L_2)}\ldots D^{\dagger}\}^{(L)}|0>, \quad (2.13)$$

where α denotes a set of intermediate angular momenta L_1, L_2,...,
$L_{\frac{n}{2}-1}$. This method of construction of states is somewhat related
to that used in ref. 6.

The subspace of the full shell model space spanned by the
states (2.13) will be called the D-subspace. Using the states
(2.13) and the operator S_+ we can construct additional states as

$$|j^n(S^{(n-v)/2}D^{v/2}),\alpha,L,M\rangle$$

$$= \sqrt{\frac{(\Omega-\frac{1}{2}(n+v))!}{((n-v)/2)!(\Omega-v)!}}\ (S_+)^{(n-v)/2}|j^v(D^{v/2}),\alpha,L,M\rangle. \quad (2.14)$$

The subspace spanned by the states (2.14) will be called the S-D
subspace in the following. The importance of the S-D subspace is
that it can be put in a one-to-one correspondence with the states of
the interacting boson model. To this correspondence (or mapping)
we now turn.

§2.3. Mapping of the S-D subspace onto a boson space

In the interacting boson model[7] a boson space was constructed
by using boson creation and annihilation operators with $L = 2$ and
$L = 0$. These operators satisfy the usual boson commutation relations

$$[d_m,d_{m'}^\dagger] = \delta_{mm'}\ ,\qquad\qquad [d_m^\dagger,d_{m'}^\dagger] = [d_m,d_{m'}] = 0,$$

$$[s,s^\dagger] = 1,\qquad [s^\dagger,s^\dagger] = [s,s] = 0,\qquad\qquad\qquad (2.15)$$

$$[s,d_m^\dagger] = [s^\dagger,d_m] = [s^\dagger,d_m^\dagger] = [s,d_m] = 0.$$

For convenience we introduce a modified operator $\tilde{d}_m = (-)^{2-m}d_{-m}$.
By repeated actions of the operators $d^\dagger$ and $s^\dagger$ on the boson
vacuum $|0\rangle$, one can construct a boson space, which we denote by s-d.
This construction can be done in two steps as in the previous sub-
sect. 2.2. First one generates states of n_d d-bosons

$$|d^{n_d},\alpha,L,M) = \mathcal{N}_{n_d\alpha}^{-1}[\ldots[d^\dagger d^\dagger]^{(L_1)}d^\dagger]^{(L_2)}\ldots d^\dagger]_M^{(L)}|0\rangle. \quad (2.16)$$

These states can be classified using the group chain $SU(5)\supset O^+(5)\supset$
$O^+(3)$ and labelled by the qauntum numbers n_d, v_d, n_Δ, L and M [ref.

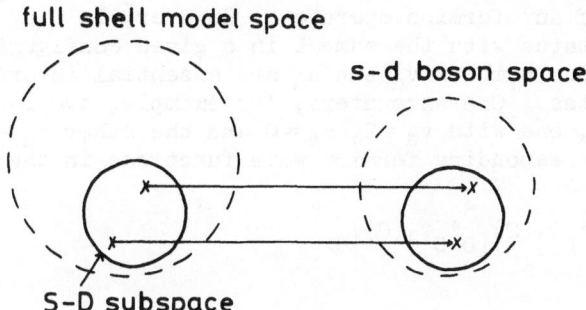

full shell model space

s-d boson space

S-D subspace

Fig. 1. Procedure in which the boson space is introduced. The full
        shell model space is truncated to the (collective) S-D sub-
        space, and the subspace is mapped onto the s-d boson space.

8]. Here n_d is the number of d-bosons, v_d is the d-boson seniority
and n_Δ is an additional quantum number which is intuitively interpreted as the number of d-boson triplets coupled to zero angular
momentum. In the second step, one acts with the operators $s^\dagger$ on the
states (2.16) thus generating states with $N = n_d + n_s$ bosons

$$|s^{n_s}d^{n_d},\alpha,L,M) = \mathcal{N}_{n_s}^{-1}(s^\dagger)^{n_s}|d^{n_d},\alpha,L,M). \qquad (2.17)$$

One can now see that, because the commutator $[D_m^\dagger, D_{m'}^\dagger]$ vanishes
when acting on a highest seniority state

$$[D_m^\dagger, D_{m'}^\dagger]|j^v, v, \alpha, L, M\rangle = 0, \qquad (2.18)$$

the states of (2.13) are symmetric under interchange of two of the
$D^\dagger$ as in eq. (2.16). This suggests that one can map the states
(2.13) onto the boson states (2.16) and similarly the states (2.14)
onto the boson states (2.17) (see. Fig. 1).
    In order to preserve the classification scheme $SU(5) \supset O^+(5) \supset$
$O^+(3)$, particular care must be taken in the mapping from the states
(2.13) onto the states (2.16). If there is only one state with a
given L in a d^{n_d} configuration no particular care is necessary.
In fact, in such a case, all different maps of coupling the angular

momenta in (2.13) and (2.16) end up with the same states, and one
can directly assign a quantum number "v_d" and "n_Λ" to the fermion
state which corresponds to the boson state carrying v_d and n_Λ.
Here we have used " ", because the numbers "v_d" and "n_Λ" are not
eigenvalues of any fermion operator. However, in many cases, there
are several states with the same L in a given configuration d^{nd}.
Here the quantum numbers v_d and n_Λ are essential in order to distinguish the states. One encounters, for example, two L = 2 states with
four d-bosons, one with $v_d = 2$, $n_\Lambda = 0$ and the other $v_d = 4$, $n_\Lambda = 1$.
One of the corresponding fermion wave functions in the D-subspace is

$$\frac{1}{\mathcal{N}_F} [D^\dagger D^\dagger]^{(2)} [D^\dagger D^\dagger]^{(0)} |0\rangle \qquad (2.19)$$

where $\mathcal{N}_F$ is the normalization constant. The state (2.19) is mapped
onto a boson state

$$|d^4, v_d=2, n_\Lambda=0, L=2) = \frac{1}{\mathcal{N}_B} [a^\dagger a^\dagger]^{(2)} [a^\dagger a^\dagger]^{(0)} |0) \qquad (2.20)$$

where $\mathcal{N}_B$ is the normalization constant. We then give the quantum
numbers "$v_d = 2$" and "$n_\Lambda = 0$" to the state (2.19). Namely the meaning
of the d-boson seniority v_d is preserved in the D-subspace. The
other state of $|j^8(D^4), L = 2\rangle$ which is obtained by orthogonalization
to the state (2.19) is labelled by the quantum numbers "$v_d = 4$" and
"$n_\Lambda = 1$". We thus classified all states of L = 2 with four D-pairs.

We can easily generalize the procedure described above to more
complicated states and classify all states constructed by using the
$D^\dagger$ operators in terms of the set of quantum numbers "v_d", "n_Λ", L
and M. Once the wave functions of the D-subspace have been constructed, one can use (2.14) to construct those of the S-D subspace

$$|j^n(D^{v/2}), "v_d", "n_\Lambda", L, M\rangle$$

$$= \sqrt{\frac{(\Omega - \frac{1}{2}(n+v))!}{((n-v)/2)!(\Omega-v)!}} \; (S_+)^{(n-v)/2}$$

$$\times |j^v(D^{v/2}), "v_d", "n_\Lambda", L, M\rangle. \qquad (2.21)$$

These states have a one-to-one correspondence with the boson states

$$|s^{n_s} d^{n_d}, v_d, n_\Delta, L, M\rangle = \frac{1}{\sqrt{n_s!}} (s\dagger)^{n_s} |d^{n_d}, v_d, n_\Delta, L, M\rangle. \qquad (2.22)$$

Thus we have established a mapping from the fermion states (2.21) to the boson states (2.22) (see Fig. 1). The mapping is such that

$$n_d = \frac{1}{2} v \qquad (2.23)$$

and the total number of bosons $n^B = n_s + n_d$ is equal to the number of fermion pairs

$$n^B = \frac{1}{2} n. \qquad (2.24)$$

III.   The boson image of fermion operators

The difficulty of the original many-body problem have been considerably reduced by going from the full shell model space to the S-D subspace. The calculation of the matrix elements in the S-D fermion subspace is, however, still rather cumbersome especially for large particle number. We want therefore to take advantage of the mapping from the S-D subspace onto the s-d boson space.

In principle, for any fermion operator in the S-D subspace, one can construct a boson image in the s-d space which reproduces exactly the results in the fermion subspace. In order to do this we have to introduce N-body boson operators even if the fermion operator is only one and/or two body. If one has to do this, the introduction of the boson space becomes superfluous. We will show, however, that the zeroth order approximation in the boson space is usually a good one. Here by m-th order of approximation to a k-body fermion operator we mean a k'-body operator, with $m = k - k'$. For example, the zeroth order approximation to a one-body fermion operator (such as the electromagnetic transition operators) is a one-body boson operator, while the zeroth order approximation to a two-body fermion operator (such as two-body interaction) is a two-body boson operator, which generally contains also one-body piece.

We have seen in the previous section the similarity between the s-d boson space and the S-D subspace. There will be also found the similarity between the Racah seniority reduction formulas[9] and the reduction formulas for s-bosons. Because of these similarities the

zeroth order approximation in the boson space is usally sufficient
to describe fermion operators in the S-D subspace. In order to show
this point, which is a key point of our scheme, and to illustrate
the procedure by means of which the boson image of a fermion operator
is constructed, we consider the case of the quadrupole operator $U^{(2)}$
in eq. (2.3). The reduction formula of the seniority changing matrix
elements of $U^{(2)}$ is

$$
<j^n, v, \alpha, L \| U^{(2)} \| j^n, v-2, \alpha', L'>
$$

$$
= \sqrt{\frac{n-v+2}{2} \times \frac{(2\Omega-n-v+2)}{2(\Omega-v+1)}}
$$

$$
\times <j^v, v, \alpha, L \| U^{(2)} \| j^v, v-2, \alpha', L'>. \tag{3.1}
$$

Because the states in the S-D subspace (2.14) have definite seniority
we can write

$$
<j^n(D^{v/2}), "v_d", "n_\Delta", L \| U^{(2)} \| j^n(D^{(v/2)-1}), "v_d'", "n_\Delta'", L'>
$$

$$
= \sqrt{\frac{n-v+2}{2} \times \frac{(2\Omega-n-v+2)}{2(\Omega-v+1)}}
$$

$$
\times <j^v(D^{v/2}), "v_d", "n_\Delta", L \| U^{(2)} \| j^v(D^{(v/2)-1}), "v_d'", "n_\Delta'", L'>,
$$

$$
\tag{3.2}
$$

where we have used explicitly the quantum numbers $"v_d"$, $"n_\Delta"$ instead
of α, and considered only $n \leq \Omega$.

The right hand side of eq. (3.2) contains two factors. The
second one $\sqrt{(2\Omega-n-v+2)/2(\Omega-v+1)}$ has v both in the numerator and the
denominator. Since the factor is insensitive to v especially for
$n \sim \Omega$, one can approximate it by a constant $\sqrt{(2\Omega-n)/2(\Omega-1)}$. The
first factor $\sqrt{(n-v+2)/2}$, however, can not be approximated by $\sqrt{n/2}$,
since v can be easily comparable to n. On the other hand, this factor
is precisely the same as the matrix element of s between the states
$|s^{n_s+1})$ and $|s^{n_s})$, once the identification $n_d = v/2$, $n^B = n/2$,
$n_s = (n-v)/2$ has been made,

$$
(s^{n_s} | s | s^{n_s+1}) = \sqrt{n_s+1} = \sqrt{\frac{1}{2}n - \frac{1}{2}v + 1}. \tag{3.3}
$$

Since the quadrupole operator $U^{(2)}$ changes an S-pair to a D-pair in this case, the operator is mapped in zeroth order onto a boson operator $\overset{0}{\alpha}_2 d^\dagger s$ where $\overset{0}{\alpha}_2$ is obtained by imposing

$$<j^n(s^{\frac{n}{2}-1}D),L=2\| U^{(2)} \|j^n(s^{\frac{n}{2}}),L=0>$$

$$= (s^{\frac{n}{2}-1}d, L=2\|\overset{0}{\alpha}_2 d^\dagger s\| s^{\frac{n}{2}}, L=0). \tag{3.4}$$

Substituting $v=2$ in eq. (3.2), we obtain

$$\overset{0}{\alpha}_2 = \sqrt{\frac{2\Omega-n}{2(\Omega-1)}} \times \frac{<j^2(D),L=2\| U^{(2)} \|j^2(S),L=0>}{(d,L=2\| d^\dagger s \|s,L=0)}. \tag{3.5}$$

The s-boson in eq. (3.3) represents the effect that the particle number is finite and conserved. The factor $\sqrt{(2\Omega-n)/2(\Omega-1)}$ in eq. (3.5) also represents the Pauli blocking effect, which is related to the finite size of the shell rather than the conservation of the particle number. The factor $\sqrt{(2\Omega-n)/2(\Omega-1)}$ is considered in the BCS scheme by the u- and v-factors, while the v-dependence in the other one $\sqrt{(n-v+2)/2}$ is not.

We obtain the coefficient $\overset{0}{\beta}_2$ of $[d^\dagger \tilde{d}]^{(2)}$ term in the zeroth order boson image of $U^{(2)}$ as (see ref. 4 or 5),

$$\overset{0}{\beta}_2 = \frac{\Omega-n}{\Omega-2} \times \frac{<j^2(D),L=2\| U^{(2)} \|j^2(D),L=2>}{(d,L=2\| [d^\dagger \tilde{d}]^{(2)} \| d,L=2)}. \tag{3.6}$$

Combining eqs. (3.5) and (3.6), we obtain the complete zeroth order image $Q^{(2)}$ of $U^{(2)}$ as

$$Q^{(2)} = \sqrt{\frac{2\Omega-n}{2\Omega-2}} \times \frac{<j^2(D)\| U^{(2)} \|j^2(S)>}{(d\| d^\dagger s \|s)} (d^\dagger s + \tilde{d} s^\dagger)$$

$$+ \frac{\Omega-n}{\Omega-2} \times \frac{<j^2(D)\| U^{(2)} \|j^2(D)>}{(d \| [d^\dagger \tilde{d}]^{(2)} \| d)} [d^\dagger \tilde{d}]^{(2)}. \tag{3.7}$$

In this expression we approximated the reduction factors, which are functions of v ($=2n_d$), by constants. If exact reduction factors are given (like in a single j-orbit), the v-dependence can be explicitly brought in the boson image if one wishes, though this does not necessarily mean improvement in our approximation.[3,4]

Although we discussed only about the quadrupole operator $U^{(2)}$, other fermion operators, for instance the Hamiltonian, are treated essentially in the same way.[3,4]

IV. Comparison with shell model calculations in a single j-orbit

Once the zeroth order boson image is constructed, it should be investigated now to what extent the image reproduces the corresponding matrix elements of states with many fermions.

Seniority changing, $\Delta v = 2$ (i.e. $\Delta n_d = 1$), matrix elements of the quadrupole operator $U^{(2)}$ ($j = 29/2$) with respect to states in the D-subspace are shown in Table I, where the matrix elements are normalized by $\langle j^2(D) \| U^{(2)} \| j^2(S) \rangle$. In the same table, matrix elements of the operator $d^\dagger s$ are also shown, being normalized by $(d \| d^\dagger s_+ \| s)$ ($= \sqrt{5}$). The matrix elements of the zeroth order boson image $d^\dagger s$ follow well the fermion matrix elements. By using eqs. (3.2) and (3.7), we obtain matrix elements in a system of a fixed n, while the same conclusion is obtained.

The zeroth order image of two-body interactions is constructed in the essentially same way as one-body operators. In Table II we compare matrix elements of the delta function interaction,

$$V_\delta = \delta(\vec{x} - \vec{x}') , \qquad (4.1)$$

in $j = 23/2$ orbit with those of its zeroth order boson image V^B ;

$$V^B = E_d n_d + \frac{1}{2} \sum_L E_L ([d^\dagger d^\dagger]^{(L)} [\tilde{d} \tilde{d}]^{(L)}) \qquad (4.2)$$

where E_d and E_L's are constants. Agreement between fermion matrix elements and boson ones is almost perfect. There is no need of three- or four-body boson-boson interaction. Because of the relation

$$\langle j^n, v, \alpha, L | V_\delta | j^n, v, \alpha', L \rangle$$

$$= \langle j^v, v, \alpha, L | V_\delta | j^v, v, \alpha', L \rangle + \frac{1}{2}(n-v) \langle j^2; 0^+ | V_\delta | j^2; 0^+ \rangle \qquad (4.3)$$

where α and α' are additional quantum numbers, the quality of

Table I. Comparison between seniority changing fermion matrix elements of the operator $U^{(2)}$ (in j = 29/2) and matrix elements of its zeroth order boson image $d^{\dagger}s$. The quantities v_d ($v_d{}'$) and L (L') are the O(5) seniority and the angular momentum of the bra- (ket-) vector. The definition of the fermion matrix elements is in eq. (3.2) with n = v, while they are normalized by the matrix element of n = v = 2. The boson matrix elements are defined in the same way.

v	L'	L	$v_d{}'$	v_d	fermion m. e.	boson m. e.
2	0	2	0	1	1	1
4	2	4	1	2	1.92	1.90
		2	1	2	1.43	1.41
		0	1	0	.55	.64
6	4	6	2	3	2.88	2.79
		4	2	3	1.64	1.60
		3	2	3	−1.13	−1.09
		2	2	1	.83	1.02
	2	4	2	3	1.73	1.68
		3	2	3	1.79	1.74
		2	2	1	.62	.76
		0	2	3	.81	.77
	0	2	0	1	1.14	1.19
8	6	8	3	4	3.89	3.69
		6	3	4	1.90	1.82
		5	3	4	−1.53	−1.45
		4	3	2	1.00	1.31
		4	3	4	.29	.28
	4	6	3	4	2.79	2.66
		5	3	4	1.54	1.45
		4	3	2	.58	.76
		4	3	4	−1.97	−1.81
		2	3	2	.61	.80
		2	3	4	.89	.83
	3	5	3	4	2.27	2.16
		4	3	2	.43	.51
		4	3	4	2.06	1.95
		2	3	2	−.65	−.82
		2	3	4	1.45	1.35
	2	4	1	2	2.09	2.16
		4	1	4	.0	.0
		2	1	2	1.57	1.60
		2	1	4	.0	.0
		0	1	0	.75	.89
	0	2	3	2	.27	.37
		2	3	4	1.29	1.21

Table II. Comparison between matrix elements in $j = 23/2$ orbit of
the fermion delta function interaction in eq. (4.1) and
the corresponding boson ones. The fermion matrix elements
are with respect to states in the D-subspace and are
normalized by $<j^2(D)|V_\delta|j^2(D)>$. The zeroth order boson
Hamiltonian is given in eq. (4.2). See the caption of
Table I.

v	L	v_d	v_d'	fermion m. e.	boson m. e.
2	2	1	1	1	E_d (= 1)
4	4	2	2	2.49	$2\,E_d + E_4$
4	2	2	2	2.42	$2\,E_d + E_2$
4	0	0	0	2.24	$2\,E_d + E_0$
6	6	3	3	4.48	4.49
	4	3	3	4.38	4.38
	3	3	3	4.35	4.34
	2	1	1	4.11	4.10
	0	3	3	4.33	4.28
8	8	4	4	6.96	6.98
	6	4	4	6.81	6.84
	5	4	4	6.79	6.78
	4	2	2	6.48	6.45
	4	4	4	6.73	6.73
	4	2	4	0.01	.0
	2	2	2	6.42	6.38
	2	4	4	6.73	6.65
	2	2	4	.0	.0
	0	0	0	6.27	6.20

Table III. Occupation probabilities in the S-D subspace of exact
shell model wave functions.

0_g^+	99.9 %	2_1^+	99.1 %	4_1^+	3.8 %
0_1^+	96.7	2_2^+	92.6	4_2^+	91.1
0_2^+	7.1	2_3^+	5.5	4_3^+	5.0

agreement for any n is the same as that in Table II.   For the pairing interaction the agreement is indeed perfect.[3,4]

We have investigated numerically other matrix elements[3,4] and
concluded that the zeroth order boson image already provides a
rather good approximation to the fermion problem in the S-D subspace.   Better approximation can be obtained by going to higher
orders, though this does not seem to be necessary in most practical
cases.

In Fig. 2 is shown comparison among diagonalizations in the
full shell model space, in the S-D subspace and in the s-d boson
space.[3,4] Here only neutron bosons are present and the Hamiltonian
consists of pairing force and (weak) quadrupole-quadrupole force.
States with an asterisk are outside the S-D subspace.   The boson
Hamiltonian is the zeroth order image of the fermion one.   In
Table III the occupation probabilities in the S-D subspace are also
presented.   One finds that states in the S-D subspace are well
separated from the rest and that the energies of the collective
states are well reproduced by the boson Hamiltonian, though the
calculation is very simple and schematic.

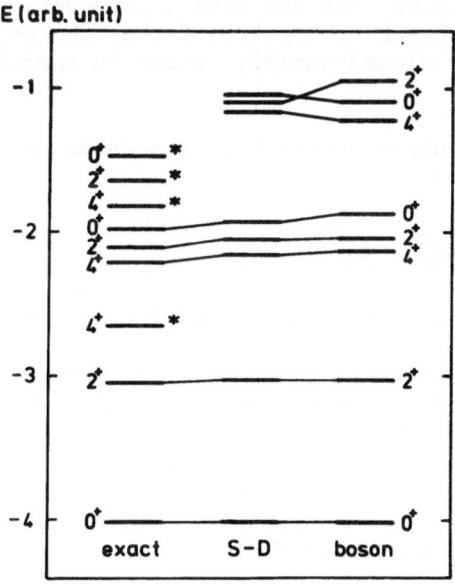

Fig. 2   Comparison among the calculated energies in the full shell
         model space of j=23/2 (exact), in the S-D subspace (S-D) and
         in the s-d boson space (boson). The three low lying states
         are calculated in each case. States with the asterisk are
         outside the S-D subspace (i.e. intruders).

V .   Spectra in a single j-orbit

We investigate what spectra are obtained by the diagonalizations of the zeroth order boson Hamiltonian H^B which is constructed as an approximation to the shell model.

Correspondingly to proton pairs and neutron pairs in nuclei, the proton s_π- and d_π-bosons and the neutron s_ν- and d_ν-bosons are introduced. The boson Hamiltonian is written as

$$H^B = H^B_\pi + H^B_\nu + H^B_{\pi\nu} \qquad\qquad (5.1)$$

where H_π, H_ν and $H_{\pi\nu}$ are the proton, the neutron and the proton-neutron boson Hamiltonians.

An example of results of such a diagonalization is shown in Fig. 3. A single $j = 31/2$ orbit, which is equivalent in a sense to the major shell 50-82 with degenerated orbits, is assumed for both protons and neutrons. Six protons ($n_\pi = 6$) are placed, while the number of neutrons is varied from 0 to 32. The boson Hamiltonian H^B_π and H^B_ν are calculated for a delta function interaction and with strength adjusted so as to produce a 0^+_g-2^+_1 separation of 1.4 MeV in single closed nuclei. The delta function interaction satisfies the general properties that the interaction between identical nucleons does not break seniority and that the separation is (almost) constant in single closed nuclei. There is therefore no term such as $d^\dagger_\pi d^\dagger_\pi s_\pi s_\pi$, etc.

The proton-neutron interaction was taken as $-f_{\pi\nu}(U^{(2)}_\pi U^{(2)}_\nu)$, which gives rise to

$$H^B_{\pi\nu} = - f_{\pi\nu} (Q^{(2)}_\pi Q^{(2)}_\nu) \qquad\qquad (5.2)$$

where $Q^{(2)}$ is given in eq. (3.7), and $f_{\pi\nu}$ is a fixed constant.

Various situations of the collective motion appear in Fig. 3 as a consequence of "the approximated shell model calculations" for various particle numbers ; anharmonic vibrator (boson SU(5) picture [ref. 8]) at $n_\nu \sim 4$, axial rotor (boson SU(3) picture [ref. 10]) at $n_\nu = 10 \sim 14$, and triaxial or γ-unstable rotor (boson O(6) picture [ref. 11]) at $n_\nu \sim 28$. In addition to these limiting cases, the intermediate situations also follow general trends observed in real nuclei.

Spectra are not symmetric with respect to the reflection at $n_\nu \sim 16$ in Fig. 3. This is because the coefficient of $[d^\dagger d]^{(2)}$ term of $Q^{(2)}_\nu$ in eq. (5.2) changes its sign at the middle of shell (see eq. (3.7)). The Hamiltonian (5.1) is (approximately) symmet-

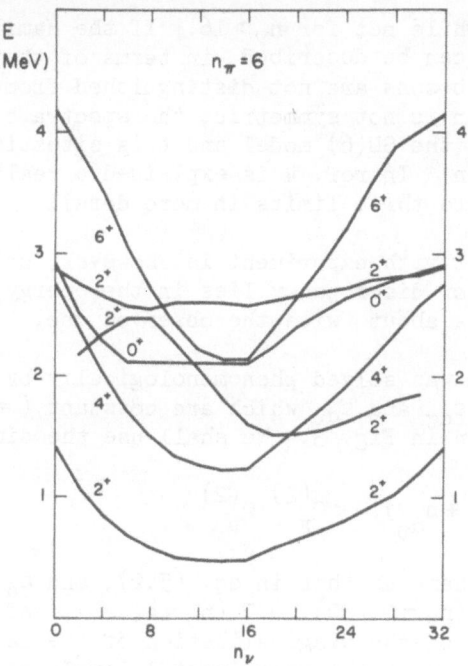

Fig. 3   Energy spectra of even-even nuclei, for fixed proton number,
n_π=6, and varying neutron number, 0∿32, in a single j=31/2
approximation. For n_ν>16, neutrons are treated as holes.

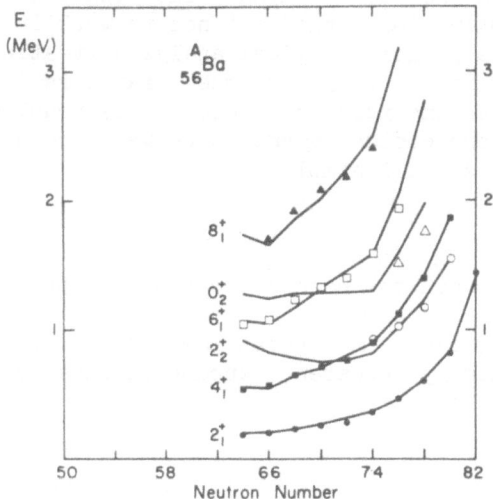

Fig. 4   Calculated energy spectra in the Ba isotopes using the
Hamiltonian (5.3). The decreased single d-boson energies
are used. The points, circles, squares and triangles are
the experimental values.

ric for $n_\nu < 16$, while not for $n_\nu > 16$. If the Hamiltonian is symmetric, the spectra can be described in terms of the SU(6) boson model
where the proton bosons are not distinguished from the neutron ones.
If the Hamiltonian is not symmetric, the spectra can be no longer
well described by the SU(6) model and this situation occurs mainly
in the O(6) region. In ref. 4 is explained a realization of the
SU(6) model and its three limits in more detail.

The agreement with experiment is, however, only qualitative in
Fig. 3. The major discrepancy lies in the energy scale ; the
theoretical one is about twice the observed one.

This problem was solved phenomenologically by reducing single
d-boson energies ε_{d_π} and ε_{d_ν} which are constant ($= 1.4$ MeV) in the
calculations shown in Fig. 3. We shall use the simple Hamiltonian

$$H^B = \varepsilon_d (n_{d_\pi} + n_{d_\nu}) - \kappa Q_\pi^{(2)} \cdot Q_\nu^{(2)} \qquad (5.3)$$

where $H^B_{\pi\nu}$ is the same as that in eq. (5.2), and ε_d (MeV) = 1.4
($n_\nu = 0,\ 32$), 0.9 ($n_\nu = 2$, 30), 0.7 ($n_\nu = 4$, 28), and 0.5 ($6 \le n_\nu \le 26$).
In Fig. 4 results of the diagonalization of the Hamiltonian (5.3)
are shown together with the experimental levels of the Ba isotopes
($n_\pi = 6$). The theoretical levels can be now compared quantitatively
with the experimental ones. E2 transition probabilities and so on
were also explained.

Investigations of other nuclei suggested that the reduction of
ε_d is required in the whole region. The single j-orbit (or degenerate orbit) approximation turned out not to work in general cases.
We then started a phenomenological analysis treating coefficients
in $Q_\pi^{(2)}$ and $Q_\nu^{(2)}$ in eq. (5.3) as free parameters in addition to ε_d.
Results of such an analysis were reported by Scholten in this work-
shop, and the phenomenology seems to be very useful and general,
while the model is still simple.

VI. Relation between the bosons and the structure of real nuclei

In realistic cases the operators S_+ in eq. (2.5) and $D^\dagger$ in eq.
(2.12) are defined by (coherent) superpositions over many j-orbits ;

$$S_+ = \sum_j \alpha_j S_{+j} \qquad (6.1)$$

$$D^\dagger = P \sum_{(ij)} \beta_{ij} A^{\dagger(2)}(ij) \qquad (6.2)$$

where α_j and β_{ij} are amplitudes and P is an appropriate operator
like P in eq. (2.11). The amplitudes should be determined so as to
make the boson approximation good. The operator S in eq. (6.1)
creates the Cooper pair in nuclei and corresponds to the S_+ operator
in Talmi's scheme[12]. The gross structure of the operator $D^\dagger$ is
determined by the Tamm-Dankoff calculation in the v=2 (or two quasiparticle) space. Once amplitudes α_j and β_{ij} are obtained, one can
calculate the zeroth order boson Hamiltonian by using a seniority
scheme in ref. 13 similarly to the cases in a single j-orbit.
Detailed discussions should be found in ref. 4.

In such a calculation the single d-boson energies ε_{d_π} and ε_{d_ν}
are still large. We then evaluate coupling effects between states
in the S-D subspace and those outside the subspace through the
Feshbach method[14]. The single d-boson energies ε_{d_π} and ε_{d_ν} are
decreased by the coupling more than the single s-boson energies.
An account of the evaluation of the coupling effects is given in
ref. 4, while this subject is still being studied.

VII.   Summary

     This talk is summarized as follows

   (i)   The s-boson corresponds to the Cooper pair in nuclei, and it
carries the effect of the particle number conservation. The meaning
and usefulness of the s-boson are different from those of the cut-
off factor of Janssen et al [ref. 15], though the cut-off factor
has a similar effect to that of the s-boson. In refs. 16 and 17
the cut-off factor is introduced to take into account different
normalizations in the boson space and fermion one, and is not related
to the particle number conservation.

   (ii)   It was shown in a single j-orbit that an appropriate one-body
boson operator is usually sufficient for a one-body fermion operator.
The similar conclusion was obtained for two-body interactions. This
is in contrast to the fact that higher order terms are important in
the boson expansion.[18,19].

   (iii)   Our method is an approximation to the shell model. We first
truncate the full shell model space to the S-D subspace, and then map
a state in the subspace onto an s-d boson state. The S-D subspace
seems to be separated from the rest to good extent, but there is
the coupling between these two. The coupling causes the renormalization of the single d-boson energies.

   (iv)   We introduced the s_π, d_π, s_ν and d_ν bosons corresponding to
proton pairs and neutron pairs. A new version of the interacting
boson model with these four kinds of bosons was proposed. The new
version is called the proton-neutron (interacting) boson model, and

describes various situations of the nuclear collective motion more
generally than the SU(6) model. The phenomenology in terms of the
new version is very promising.

References

1. A. Arima, T. Otsuka, F. Iachello and I. Talmi, Phys. Lett.
   66B:205 (1977).
2. T. Otsuka, A. Arima, F. Iachello and I. Talmi, Phys. Lett.
   76B:139 (1978).
3. T. Otsuka, A. Arima and F. Iachello, Nucl. Phys. A309:1 (1978).
4. T. Otsuka, Doctor Thesis, University of Tokyo (1978).
5. A. K. Kerman, Ann. Phys. (N.Y.) 12:300 (1961).
6. S. G. Lie and G. Holzwarth, Phys. Rev. C12:1035 (1975).
7. A. Arima and F. Iachello, Phys. Rev. Lett. 35:1069 (1975).
8. A. Arima and F. Iachello, Ann. Phys. (N.Y.) 99:253 (1976).
9. A. de Shalit and I. Talmi, "Nuclear Shell Theory", Academic
   Press, New York (1963).
10. A. Arima and F. Iachello, Ann. Phys. (N.Y.) 111:201 (1978).
11. A. Arima and F. Iachello, Phys. Rev. Lett. 40:385 (1978).
12. I. Talmi, Nucl. Phys. A172:1 (1971).
13. T. Otsuka and A. Arima, Phys. Lett. 77B:1 (1978).
14. H. Feshbach, Ann. Phys. (N.Y.) 5:357 (1958); 19:287 (1962).
15. D. Janssen, R. V. Jolos and F. Donau, Nucl. Phys. A224:93 (1974).
16. G. Holzwarth, D. Janssen and R. V. Jolos, Nucl. Phys. A261:1
    (1976).
17. S. Iwasaki, F. Sakata and K. Takada, Prog. Theor. Phys. 57:1289
    (1977).
18. B. Sorensen, Nucl. Phys. A217:505 (1973).
19. T. Kishimoto and T. Tamura, Nucl. Phys. A270:317 (1977).

FERMION HAMILTONIANS WITH MONOPOLE AND QUADRUPOLE PAIRING

J. N. Ginocchio

Theoretical Division, Los Alamos Scientific Laboratory

University of California, Los Alamos, New Mexico 87545

INTRODUCTION

In this workshop we have seen evidence that the interacting boson model (IBM) seems to provide a unified phenomenological model of vibrational, transitional, and rotational nuclei. The underlying concept of this model is that the low-lying collective states of heavy nuclei are made up of monopole and quadrupole Bosons $s^\dagger$ and $d_\mu^\dagger$ bosons where $\mu = -2,-1,0,1,2$. These bosons can be interpreted to represent correlated pairs of valence nucleons outside the **closed core. Hence a** natural view of the IBM is that it is an approximation to the complicated shell model description of these heavy nuclei. That is, the valence nucleons move in an average field and the residual interactions produce correlated monopole, $S^\dagger$, and quadrupole, $D_\mu^\dagger$, pairs of valence nucleons. The collective low-lying states are then comprised mostly of these pairs, and the effect of the many other more complicated states is mainly to renormalize the shell model Hamiltonian and transition operators.

Given this viewpoint it is natural to ask the question: Are there shell model Hamiltonians which will have a class of eigenstates made up only of monopole and quadrupole pairs? The answer to the question is yes, and I shall discuss examples of such Hamiltonians in this talk. Of course these Hamiltonians will be model Hamiltonians in the same sense that the famous pairing Hamiltonian with degenerate single-particle energies is a model Hamiltonian. Nevertheless these models can be very instructive for understanding the microscopic structure of the **IBM and also may provide insight** on how more realistic shell model Hamiltonians may provide eigenstates made up primarily of monopole and quadrupole pairs.

DISCUSSION

The monopole and quadrupole pair operators will be linear combinations of nucleon pairs in spherical shell model orbits labeled by angular momentum j and projection m:

$$S^{\dagger} = \tfrac{1}{2} \Sigma_j \alpha_j \Sigma_m (-1)^{j-m} A^{\dagger}_{jm} A^{\dagger}_{j-m} \qquad (1a)$$

$$D^{\dagger}_{\mu} = \Sigma_{j,j'} \beta_{jj'} \left[A^{\dagger}_j A^{\dagger}_{j'} \right]^2_{\mu} \qquad (1b)$$

where $A^{\dagger}_{jm}$ creates a nucleon in orbit (j,m) and the brackets $\left[\; \right]^J_{\mu}$ mean that these operators are coupled to angular momentum J and projection μ,

$$\left[A^{\dagger}_j A^{\dagger}_{j'} \right]^J_{\mu} \equiv \Sigma_{m,m'} (jmj'm' | jj'J\mu) A^{\dagger}_{jm} A^{\dagger}_{jm'} \qquad (1c)$$

where $(jmj'm' | jj'J\mu)$ is a Clebsch–Gordon coefficient.

We consider a shell model Hamiltonian H which has a spherical average one-body field h and a two-body interaction υ between valence nucleons:

$$H = \sum_{i=1}^{n} h_i + \sum_{i<j}^{n} \upsilon_{ij} \qquad (2)$$

where n is the number of valence nucleons. We are interested in such Hamiltonians which have a class of eigenstates which are linear combination of the states made up of monopole and quadrupole nucleon pairs:

$$| N \, N_d \gamma JM \rangle = \left(S^{\dagger} \right)^{N-N_d} \left(D^{\dagger} \right)^{N_d}_{\gamma JM} | 0 \rangle \quad , \qquad (3a)$$

where

$$N = \tfrac{1}{2} n \quad , \qquad (3b)$$

and N_d is the number of quadrupole pairs. These quadrupole pairs are coupled to angular momentum J and projection M, and γ refers to any additional quantum numbers which may be necessary. We refer to the space spanned by the states in (3) by $(S^{\dagger}, D^{\dagger})^N$ for convenience.

The necessary and sufficient conditions that a shell model Hamiltonian have a class of eigenstates in the $(S^{\dagger}, D^{\dagger})^N$ space are an extension of those given by Talmi for generalized seniority[1,2]. One set of conditions is that the pair operators (1) create two-nucleon states which are eigenstates of H,

$$[\text{H},\text{s}^{\dagger}]|0\rangle = E_0 \text{s}^{\dagger}|0\rangle \tag{4a}$$

$$[\text{H},\text{D}_\mu^{\dagger}]|0\rangle = E_2 \text{D}_\mu^{\dagger}|0\rangle \tag{4b}$$

where $|0\rangle$ represents the core of A-n nucleons, where A is the total number of nucleons in the nucleus.

Another set of conditions is that the double commentators of the pair operators give back the pair operators:

$$\left[[\text{H},\text{s}^{\dagger}],\text{s}^{\dagger}\right] = (00|\Delta|00)\text{s}^{\dagger}\text{s}^{\dagger} + (00|\Delta|22)\ \text{D}^{\dagger}\text{D}^{\dagger} \tag{5a}$$

$$\left[[\text{H},\text{s}^{\dagger}],\text{D}_\mu^{\dagger}\right] = (02|\Delta|02)\text{s}^{\dagger}\text{D}_\mu^{\dagger} + (02|\Delta|22)\left[\text{D}^{\dagger}\text{D}^{\dagger}\right]_\mu^2 \tag{5b}$$

$$\left[[\text{H},\text{D}_\mu^{\dagger}],\text{D}_{\mu'}^{\dagger}\right] = \delta_{\mu,-\mu'}\,(-1)^\mu(22|\Delta|00)\text{s}^{\dagger}\text{s}^{\dagger}$$

$$+ 2\,(2\mu2\mu'|22\ 2,\mu+\mu')(22|\Delta|02)\text{s}^{\dagger}\text{D}_{\mu+\mu'}^{\dagger} \tag{5c}$$

$$+ \sum_{J=0,2,4}(2\mu2\mu'|22J,\mu+\mu').(22|\Delta_J|22)\left[\text{D}^{\dagger}\text{D}^{\dagger}\right]_{\mu+\mu'}^{J}$$

These conditions are very restrictive. One way to satisfy them is by means of group theory. Since the pair creation operators span a six-dimensional space we look for groups which have a six-dimensional irreducible representation which contain states with angular momentum zero and two. There are only three such groups, SU_3, SO_6 and SU_6. If we then include single-particle levels which fit into representations of one of these groups, we can then construct an H which is invariant with respect to the group, and hence the conditions (4) and (5) must be satisfied. However of the three groups, only one has representations which have angular momenta corresponding to single-nucleon angular momenta. This group is the SO_6 group. In a recent paper[3] we considered a particular example of this method which corresponded to valence nucleons filling the (1p, 0f) major shell. Other examples corresponding to other shells can be constructed as well.

This method restricts the possible single-particle angular momenta that are allowed. There is another method to obtain solutions which is less restrictive. We will not go into the details of this method, which will be explained in a forthcoming paper, but just give the solution.

For this solution the single-particle orbitals are degenerate in energy, as is the case for the exactly soluble pairing model. This feature may not be a severe handicap because for many valence nucleons the two-body interactions will be more important than the single-particle energies.

Furthermore for this solution the single-particle orbitals will come in groups with all j between the limits,

$$k + \frac{3}{2} \geq j \geq |k - \frac{3}{2}|, \tag{6}$$

being necessary. Here k is any integer. We note that for k=1, the orbitals are those of the (1s,0d) major shell, and for k=2, of the (1p,0f) major shell. Different combinations of orbitals may be taken, as long as the k's differ by four, so that a particular angular momentum does not occur more than once. We don't feel that this grouping has any deep physical meaning but is just a peculiarity of the model. What is significant is that, except for the trivial case of k=0 (j=3/2 only), many spherical orbitals are needed to have a decoupling of the $(S^\dagger, D^\dagger)^N$ space from the other complicated states.

A set of Hamiltonians which are solutions of equations (4) and (5) and hence have this decoupling feature is given by,

$$H = \overline{E}_o \, S^\dagger S + \overline{E}_2 D^\dagger \tilde{D} + \frac{1}{4} \sum_{r=0}^{3} b_r \, P^r \cdot P^r, \tag{7}$$

where $S^\dagger$ is the usual pairing mode,

$$S^\dagger = (4\Omega)^{-\frac{1}{2}} \sum_{jm} (-1)^{j-m} A_{jm}^\dagger A_{j-m}^\dagger \tag{8a}$$

the quadrupole pairing mode is,

$$D_\mu^\dagger = \sum_{j,j'} (-1)^{k+\frac{3}{2}+j'} \left[\frac{(2j+1)(2j'+1)}{\Omega} \right]^{\frac{1}{2}} \begin{Bmatrix} \frac{3}{2} & \frac{3}{2} & 2 \\ j & j' & k \end{Bmatrix} \left[A_j^\dagger A_{j'}^\dagger \right]_\mu^2, \tag{8b}$$

where $\begin{Bmatrix} j_1 & j_2 & j_3 \\ j_4 & j_5 & j_6 \end{Bmatrix}$ is the Wigner 6-j symbol and,

$$\tilde{D}_\mu \equiv (-1)^\mu D_{-\mu}, \tag{8c}$$

and Ω is one-half of the total occupancy of the shell

$$\Omega \equiv \frac{1}{2} \sum_j (2j+1) = 2(2k+1). \tag{8d}$$

These pairing modes are normalized so that they create a normalized two-nucleon state,

$$\langle o | SS^\dagger | 0 \rangle = \langle o | D_\mu D_\mu^\dagger | 0 \rangle = 1 \tag{8e}$$

The multipole operators are given by

$$P_q^r = 2 \sum_{j,j'} \left[(2j+1)(2j'+1) \right]^{\frac{1}{2}} (-1)^{k+r+\frac{3}{2}+j'} \begin{Bmatrix} \frac{3}{2} & \frac{3}{2} & r \\ j & j' & k \end{Bmatrix} \left[A_j^\dagger \tilde{A}_{j'} \right]_q^r \tag{9a}$$

with

$$\tilde{A}_{jm} \equiv (-1)^{j+m} A_{j-m} \tag{9b}$$

The integer k is the same as mentioned in (6). The multipole operator of zero rank is just the fermion valence number operator

$$P^0 = n \tag{9c}$$

and, for the states within the $(S^\dagger, D^\dagger)^N$ space, the dipole operator is proportional to the angular momentum operator.

Because of the multipole interactions the energy difference for the two nucleon spectrum is given by

$$E_2 - E_0 = \overline{E}_2 - \overline{E}_0 + 4(b_3 - b_2) + \frac{6(b_1 - b_2)}{5} \,. \tag{10}$$

These solutions have a very important feature. The set of states given in (3) have a one-to-one correspondence to the boson basis,

$$\left| N\ N_d \gamma JM \right\rangle_B = (\delta^\dagger)^{N-N_d} (d^\dagger)^{N_d}_{\gamma JM} \left| 0 \right\rangle_B \tag{11}$$

and all the states are Pauli allowed for $N_d \leq \frac{\Omega}{2}$. For $N_d > \frac{\Omega}{2}$, instead of valence particles, the one-to-one correspondence is between valence holes. This is consistent with the assumption of the IBM that for the number of valence nucleons greater than the half-filled shell the bosons represent the creation of a pair of holes in the full shell.

The eigenvalues of the Hamiltonian in (7) depend on the parameters of the Hamiltonian and in general need to be solved for numerically. However for special values of the parameters there are analytical expressions for the eigenvalues. One interesting case occurs when the quadrupole pairing and quadrupole-quadrupole interaction are related by

$$\overline{E}_2 = \Omega b_2 \tag{12a}$$

In that limit nucleon seniority, v, is a good quantum number and the eigenspectrum for states in the $(S^\dagger, D^\dagger)^N$ space is that of an anharmonic vibrator. The eigenspectrum is given by

$$E^*(Nv\tau JM) \equiv E(Nv\tau JM) - E(Nv=0, \tau=0, J=0)$$

$$= (E_2 - E_0) \frac{v}{4\Omega} (2\Omega + 2 - v)$$

$$+ (b_3 - b_2) \left[\tau(\tau+3) - \frac{v}{\Omega} (2\Omega + 2 - v) \right]$$

$$+ \frac{(b_1-b_3)}{5} \left[J(J+1) - \frac{3v}{2\Omega} (2\Omega+2-v) \right] \tag{12b}$$

The allowed values of seniority are limited by the total number of valence nucleons:

$$v=0,2,\ldots, 2N ; N \leq \tfrac{1}{2}\Omega \tag{13a}$$

$$v=0,2,\ldots,2(\Omega-N); N \geq \tfrac{1}{2}\Omega \tag{13b}$$

The first term in the eigenvalue spectrum (12) is just the pairing interaction and gives the main vibrational spectrum which is harmonic in v for low v. For large v this energy difference between levels decreases due to a Pauli correction.

The next terms in (12) split the degeneracy of the harmonic spectrum. The eigenvalue τ is the number of quadrupole operators which are not coupled to zero. The allowed values are

$$\tau = \tfrac{1}{2}v, \tfrac{1}{2}(v-4),\ldots, 1 \text{ or } 0. \tag{14}$$

For each τ the allowed angular momenta are determined by partitioning τ as

$$\tau = 3n_\Delta + \lambda \tag{15}$$

where n_Δ and λ are any positive integers. The allowed angular momenta for a given τ and n_Δ is

$$J = \lambda,\lambda+1,\ldots,2\lambda-2,2\lambda. \tag{16}$$

Hence the allowed eigenvalues are the same as given by Arima and Iachello[4] for the vibrational limit of the IBM.

In Fig. 1 we show a schematic representation of the vibrational spectrum given in (12), for $v \leq 6$ which illustrates the roles of the three terms in (12). The detailed ordering of the levels will depend on the details of the parameters and could differ from that given in Fig. 1.

Furthermore if we assume that the electric quadrupole operator is proportional to the quadrupole multipole operator given in (9a),

$$Q_q = \chi P_q^2 \tag{17}$$

then it follows that the matrix elements of Q_q between any states with the same seniority is exactly zero:

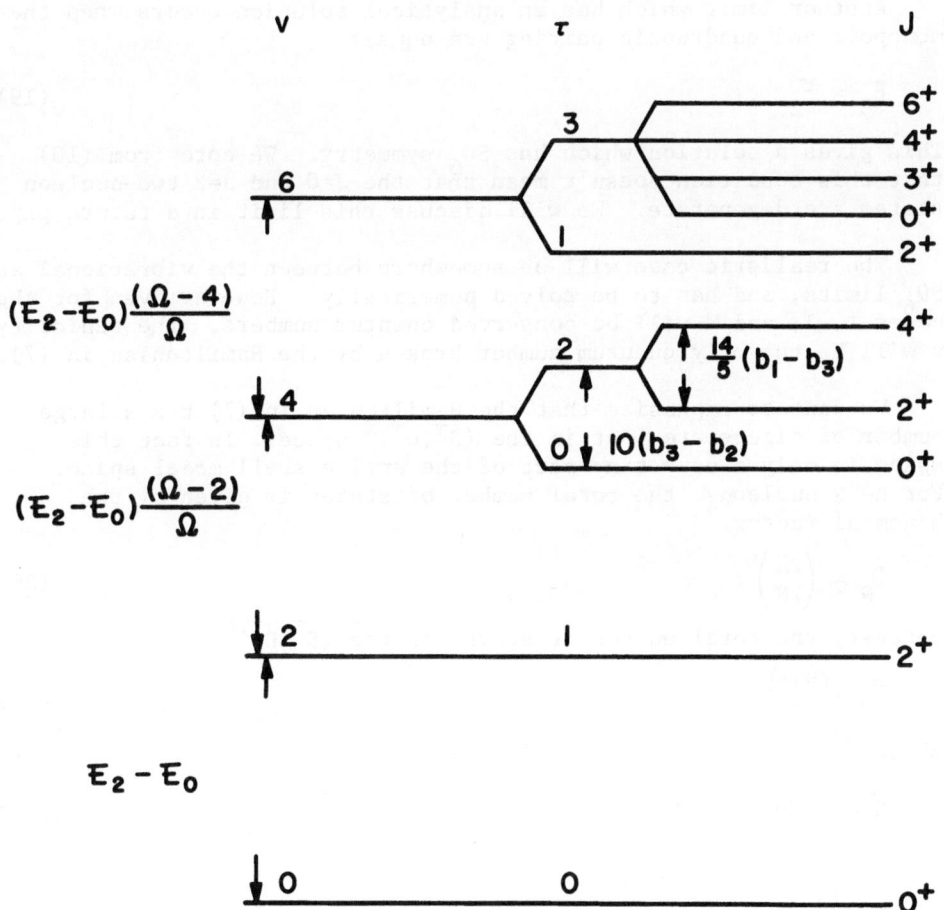

Fig. 1 A schematic representation of the vibrational spectrum of
equation (12) for the lowest seniorities (v ≤ 6).

$$(N\upsilon\tau'J'M'|Q_q|N\upsilon\tau JM) = 0 \quad . \tag{18}$$

This means that all quadrupole moments vanish and there are no BE2 transitions between states in the same seniority multiplet, only transitions between seniority multiplets. Hence this gives the selection rules of the extreme vibrational limit.

Another limit which has an analytical solution occurs when the monopole and quadrupole pairing are equal:

$$\bar{E}_0 = \bar{E}_2 \quad . \tag{19}$$

This gives a solution which has SO_6 symmetry. We note from (10) that this condition doesn't mean that the J=0 and J=2 two-nucleon states are degenerate. We will discuss this limit in a future paper.

The realistic case will be somewhere between the vibrational and SO_6 limits, and has to be solved numerically. However even for these cases τ, J, and M will be conserved quantum numbers. The seniority υ will be the only quantum number broken by the Hamiltonian in (7).

We want to emphasize that the Hamiltonian in (7) has a large number of eigenstates not in the $(S^\dagger, D^\dagger)^N$ space. In fact this space is only a very tiny part of the entire shell model space. For n=2N nucleons, the total number of states is given by the binomial factor

$$\mathcal{D}_N = \binom{2\Omega}{2N} \tag{20}$$

However, the total number of states in the $(S^\dagger, D^\dagger)^N$ space is

$$\bar{\mathcal{D}}_N = \binom{\bar{N}+5}{5} \quad . \tag{21a}$$

where

$$\bar{N} = N \text{ for } N \leq \Omega/2 \tag{21b}$$

and

$$\bar{N} = \Omega-N \text{ for } N \geq \Omega/2 \tag{21c}$$

Therefore, for $\Omega, \bar{N}$ large, the ratio is

$$\frac{\bar{\mathcal{D}}_N}{\mathcal{D}_N} \approx e^{-2\bar{N}\ln 2(\Omega-\bar{N})+\frac{11}{2}\ln\frac{\bar{N}}{5}} \tag{22a}$$

which is very small for $\bar{N}$ large, since

$$2(\Omega-\bar{N}) \geq \bar{N} \tag{22b}$$

for all values of $\overline{N}$. Hence if only the $(S^\dagger,D^\dagger)^N$ space is playing
a vital role in the collective low-lying states of real nuclei, the
IBM does provide a very substantial reduction in the complexity of
the shell model calculation of these states.

CONCLUSION

     We have shown that there exists fermion Hamiltonians which
decouple states containing monopole and quadrupole pairs from the
rest of the spectrum. These Hamiltonians have both a vibrational
(good seniority) and an SO_6 limit depending on the parameters of
the Hamiltonian.

     We have only discussed identical nucleons in this paper. Of
course the properties of the transitional and rotational nuclei
depend crucially on the interaction between protons and neutrons.
We will be able to study this aspect as well by introducing proton-
neutron interactions into the Hamiltonians of equation (7). We
can also study the effect of the coupling of the other states on
the spectra of the states in the $(S^\dagger,D^\dagger)^N$. By mapping H onto a
boson Hamiltonian H_B which is active in the $(s^\dagger,d^\dagger)^N$ boson space,
we can study the effect of the Pauli principle and the effect of
the states not in the $(S^\dagger,D^\dagger)^N$ space on the renormalization of the
boson Hamiltonian. We also will be able to study the coupling of
a single-nucleon to the $(S^\dagger,D^\dagger)^N$ space and hence study odd nuclei
as well. Thus by studying these, albeit schematic, fermion Hamiltonians we can perhaps understand the success of the phenomenological interacting boson model in describing real nuclei.

*Work supported by the Department of Energy.

REFERENCES

1. I. Talmi, Nucl. Phys. A172:1 (1971).
2. S. Shlomo and I. Talmi, Nucl. Phys. A198:81 (1972).
3. J.N. Ginocchio, Phys. Lett. 79B:173 (1978).
4. A. Arima and F. Iachello, Ann. Phys. 99:253 (1976).
5. A. Arima and F. Iachello, Phys. Rev. Lett. 40:385 (1978).
6. J.A. Cizewski, R.F. Casten, G.J. Smith, M.L. Stelts, W.R. Kane,
    H.G. Börner, and W.F. Davidson, Phys. Rev. Lett. 40:167 (1978).

SHELL MODEL TESTS OF THE INTERACTING BOSON MODEL DESCRIPTION

OF NUCLEAR COLLECTIVE MOTION*

J. B. McGrory

Oak Ridge National Laboratory

Oak Ridge, Tennessee 37830, U.S.A.

The IBM assumes that collective behavior arises from the
coupling, through the neutron-proton interaction, of the separate
low-lying states systems of valence protons and neutrons defined
with respect to a major shell closure. The eigenstates of the
proton (neutron) systems are assumed to be constructed purely
from combining two-particle "bosons" with L = 0 and L = 2 to
form many-particle states. The model is capable of handling
nuclear systems which are far beyond the domain of applicability
of any reasonably complete shell-model calculation. It is the
purpose of this report to present the results of a large shell-
model calculation of pseudo-nucleus which displays striking
collective behavior suggestive of rotational phenomena. In these
calculations, a specific and physically reasonable single-particle
structure is given to the wave functions, and an explicit two-
body residual interaction is used. An analysis of the shell
model wave functions of the eigenstates of the resulting K = 0
rotational bands offers strong support to a primary assumption
of the IBM; i.e., wave functions which describe collective behavior can be constructed from many-particle states of valence
nucleons which are constructed only from J = 0 and J = 2 two-
particle states.

The calculations here were suggested by earlier calculations
of Hecht et al.[1] The model space for the calculations reported
here includes the $0f_{5/2}$, $1p_{3/2}$, and $1p_{1/2}$ proton single-particle

*Research sponsored by the Division of Physical Research, U.S.
Department of Energy, under contract W-7405-eng-26 with the
Union Carbide Corporation.

orbits, and the $0g_{7/2}$, $1d_{5/2}$, $1d_{3/2}$, and $2s_{1/2}$ neutron single-particle orbits. The single-particle levels are taken to be degenerate. For a reasonably large number of particles, one-body spin-orbit effects should be minimal. For the residual shell-model Hamiltonian, the surface delta interaction[2] (SDI) is used with equal strengths in the p-p, p-n, and n-n systems. The SDI is known to be a useful approximation to more empirical and/or more realistic interactions. This model space is still a very large one. A truncation scheme is used here as follows. The Hamiltonian can be written in obvious notation, as

$$H = H_{pp} + H_{nn} + H_{pn}.$$

H_{pp} and H_{nn} are diagonalized exactly in the complete proton and neutron model spaces. The basis space of the truncated n-p-model is formed by coupling together a selected set of eigenstates of the neutron and proton systems; i.e.,

$$|\psi_{p,n}^{i,j}\rangle = |\psi_p^i\rangle \times |\psi_n^j\rangle.$$

The justification for selecting these particular eigenstates is discussed in some detail in Hecht et al. The reasoning is reviewed briefly here. For the SDI in the n-p basis space used here

$$\langle\psi_{p,n}^{i,j}|H_{p,n}^{SDI}|\psi_{p,n}^{\ell,m}\rangle \approx \sum_k \langle\psi_p^i|Q_p^k|\psi_p^\ell\rangle \cdot \langle\psi_n^j|Q_n^k|\psi_n^m\rangle$$

where Q_m^k is proportional to the spherical harmonic Y_m^k. It is demonstrated in Hecht et al. that for the set of neutron and proton states listed in Table I, the matrix elements of the surface harmonics, Q_m^k, between states within the set and states not in the set are very small. Thus, the eigenstates resulting from diagonalizing $H_{p,n}$ in the truncated basis we use are essentially uncoupled to exact eigenstates of H which have significant admixtures of proton or neutron states not included in our model space. In the truncation scheme used here the maximum dimension of any matrix model is 130, as compared to 45,000 in the full space.

Hecht et al. introduce the concept of a "favored pair". A favored pair is an eigenstate of the SDI of a system of two identical particles distributed in all the (degenerate) orbits of a complete oscillator shell with S = 0, L = J, and with a non-zero eigenvalue. There is only one such eigenstate for each even value of L up to the maximum possible L-value for a two-particle state in the given model space. All other two-particle eigenstates are

degenerate, with E = 0. Thus, in the (5/2,3/2,1/2) space, there
are favored pairs with J = 0, 2, and 4. In the (7/2,5/2,3/2,1/2)
space there are favored pairs with J = 0, 2, 4, and 6. Hecht
et al. show that the four-particle eigenstates listed in Table I,
and included in the truncated model space here, can be constructed
essentially completely from coupling of two favored pair states.
In the IBM model, the many-particle states are formed from J = 0
and J = 2 identical-particle pairs. In the calculations discussed here, I include J = 4 and J = 6 pairs in addition to the
J = 0 and J = 2 pairs. To the extent that the calculations reported here are a good approximation of a realistic shell-model
calculation in a many-particle space, the importance of the J = 4
and J = 6 pairs in the many-particle space reflects the adequacy
or inadequacy of the restriction to J = 0 and J = 2 pairs of the
IBM.

In Fig. 1 is shown the calculated spectrum of low-lying states
of a system (4 x 6) with six protons and four neutrons in the

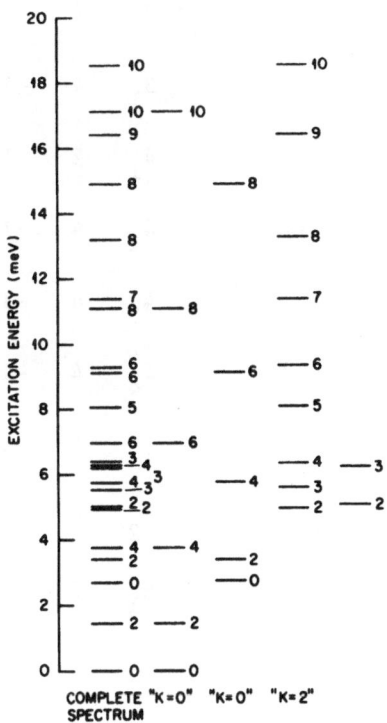

Fig. 1. Calculated Spectrum, states in the four-neutron six-
         proton system, described in the text.

Table I. Eigenstates included in truncated calculations. Column
headed ν shows the seniority of the state. E is the eigenvalue
of the state. Column headed % (2 × 2) shows percentage of given
eigenstate which is formed from coupling of J = 2 two-particle
states.

	PROTONS				NEUTRONS		
J	ν	E	% (2 × 2)	J	ν	E	% (2 × 2)
0_1	0	−18.00		0_1	0	−20.00	
0_2	6	−14.16		0_2	4	−12.82	63
2_1	2	−15.43		2_1	2	−16.67	
2_2	4	−13.96	83	2_2	4	−14.08	92
3_1	4	−12.90		2_3	4	−11.48	10
3_2	6	−11.35		3_1	4	−12.31	
4_1	2	−13.43		4_1	2	−13.94	
4_2	4	−12.73	86	4_2	4	−13.30	93
4_3	6	−11.96		4_3	4	−11.70	7
5_1	4	−11.14		4_4	4	−10.12	2
6_1	4	−10.28		5_1	4	−10.86	
6_2	6	− 9.96		5_2	4	−10.41	
				6_1	2	−11.63	
				6_2	4	−10.38	
				6_3	4	− 8.50	

model space described above. The first column in Fig. 1 shows the
spectrum of all states up to 5 MeV excitation and selected high-
spin states up to 20 MeV. B(E2) values for transitions between
these states have been calculated with total proton and neutron
charges of 1.0e, respectively. In succeeding columns in Fig. 1,
states are grouped into bands determined by observing which
states are connected by strong B(E2)-values.

   The last column in each figure shows the states not included
in the "bands". In Fig. 1 we see that a number of states in the
(4 x 6)-system can be grouped into three bands. The "ground
state" band has the level ordering and spacings very similar to
the J(J+1) spacing of a rigid rotor with K = 0. The level energies of the second and third bands strongly resemble those of
K = 2 and a K = 0 rotational bands, respectively. In Table II,
the relative B(E2)-values for transitions within the two "K = 0^{+}"
bands and the "K = 2^{+}" bands of the (4 x 6) system are shown. For
comparison, the relative B(E2)-values predicted by the usual rigid
rotor model for K = 0 and K = 2 bands are shown, i.e.,

$$B(E2) \propto \left| C_{K \; 0 \; K}^{J_i \; 2J_f} \right|^2 .$$

There is remarkable agreement between the shell model and the
rigid rotor for the K = 0_1^+ and the K = 2_1^+ band for states with
J ≤ 8. There is a band crossing at the J = 8^+ level between
the second K = 0_2^+ band and the K = 2_1^+ band. The J = 6^+ members
of the K = 0_2^+ and K = 2_1^+ bands are almost degenerate, and this
fact may be reflected in the deviations between shell model and
rigid rotor for transitions involving these states. There are
similar results in the system with four neutrons and four protons
(4 x 4), but the rotational behavior breaks down above the J = 6
state in both bands and the second K = 0^+ band is not present in
the (4 x 4) calculation. Thus, from these calculated results,
one sees that there is clear evidence of rigid-rotor behavior
appearing in the large calculations, considerably more striking
than I have seen in previous shell-model calculations. The
collective behavior is more distinct when two protons are added
to go from the 4 x 4 to the 4 x 6 system. This behavior appears
with a neutron-proton interaction (the SDI) which does not obviously enhance the possibility for rotational behavior.

   Some analysis of the wave functions of the ground-state
band of the 6 x 4 system is shown in Table III. In this table,
for each spin, all basis states admixed with coefficients greater
than 0.2 are shown, and the components are listed in decreasing
order of magnitude. In all cases, these components account for

Table II. Relative B(E2)-values in four-neutron, six-proton
systems. (The transitions are normalized to one for the first
transition in each column. Column headed S.M. gives shell model
results. Column headed R.R. gives calculated relative B(E2)'s
for rigid rotor model as discussed in the text.)

"K = 0"			"K = 0"			"K = 2"		
$J_i \rightarrow J_f$	S.M.	R.R.	$J_i \rightarrow J_f$	S.M.	R.R.	$J_i \rightarrow J_f$	S.M.	R.R.
2-0	1	1	2-0	1	1	3-2	1	1
4-2	1.48	1.43	4-2	1.34	1.43	4-2	.31	.30
6-4	1.56	1.57	6-4	1.09	1.57	4-3	.72	.75
8-6	1.42	1.65	8-6	0.61	1.65	5-3	.53	.53
10-8	1.06	1.69	10-8	0.55	1.69	5-4	.51	.53
12-10	0.5	1.72	12-10	0.24	1.72	6-4	.49	.66
						6-5	.41	.39
						7-5	.70	.74
						7-6	.34	.30
						8-6	.74	.80
						8-7	.28	.24
						9-7	.60	.83
						9-8	.18	.19
						10-8	.62	.87
						10-9	.18	.16

Table III. Principal components of ground state band wave functions in six-proton, four-neutron system. The number in each column is the observed value of the admixture of the given basis state in the eigenstate. The numbers in parentheses $(J_1 \times J_2)$ indicate the basis state $(J_p \times J_n)$, where J_p and J_n are given in Table I.

$J = 0^+$	$J = 2^+$	$J = 4^+$	$J = 6^+$
0.600 $(0_1 \times 0_1)$	0.458 $(2_1 \times 0_1)$	0.454 $(2_1 \times 2_1)$	0.401 $(2_1 \times 4_2)$
0.594 $(2_1 \times 2_1)$	0.424 $(0_1 \times 2_1)$	0.320 $(4_2 \times 0_1)$	0.391 $(4_2 \times 2_1)$
0.229 $(4_2 \times 4_2)$	0.275 $(4_2 \times 2_1)$	0.313 $(2_1 \times 2_2)$	0.328 $(4_2 \times 2_2)$
0.218 $(2_2 \times 2_2)$	0.266 $(2_1 \times 2_1)$	0.296 $(0_1 \times 4_2)$	0.296 $(2_1 \times 4_3)$
0.210 $(2_1 \times 2_2)$	0.260 $(2_1 \times 2_2)$	0.229 $(4_2 \times 2_1)$	0.203 $(2_1 \times 4_1)$
	0.255 $(2_2 \times 4_2)$	0.213 $(2_1 \times 4_2)$	0.200 $(4_2 \times 4_2)$
	0.216 $(0_1 \times 2_2)$		

$J = 8^+$	$J = 10^+$	$J = 12^+$
0.425 $(4_2 \times 4_2)$	0.476 $(4_2 \times 6_1)$	0.515 $(6_1 \times 6_1)$
0.331 $(4_2 \times 4_3)$	0.464 $(4_2 \times 6_2)$	0.491 $(6_1 \times 6_2)$
0.315 $(2_1 \times 6_2)$	0.425 $(6_2 \times 4_2)$	0.486 $(6_2 \times 6_1)$
0.314 $(2_1 \times 6_1)$	0.373 $(6_2 \times 4_3)$	0.472 $(6_2 \times 6_2)$
0.307 $(6_2 \times 2_1)$	0.209 $(6_2 \times 4_1)$	
0.278 $(6_2 \times 2_2)$		
0.258 $(4_2 \times 4_1)$		

60% or more of the wave function amplitude. For states up to the $J = 6^+$ state, these large components are dominated by coupling of $J = 0$ and $J = 2$ two-particle bosons, as assumed by the IBM. One might argue that this results from simple perturbation theory arguments; i.e., that the $J = 0^+$ and $J = 2^+$ states lie lowest in energy. However, note that besides the $J = 0_1^+$ and

$J = 2_1^+$ states, the other important states are the $J = 2_2$ state
with seniority $\nu = 4$, the $J = 4_2$ state with $\nu = 4$, and the $J = 4_3$
state with $\nu = 6$. The low-lying $J = 4_1$, $\nu = 2$ (which must be the
coupling of a $J = 0$ and a $J = 4$ pair) state plays essentially no
role in the important components of the $J = 0$, 2, 4, and 6^+ levels.
As shown in Hecht et al., these two $\nu = 4$ and one $\nu = 6$ states are
constructed essentially from the coupling of the $J = 0^+$ and
$J = 2^+$ two-particle states of the neutron and proton system. The
$J = 8$ state has important components which cannot be accounted `
for by coupling $J = 0$ and $J = 2$ two-particle states, but many
of the largest components can be so described. Note that the
agreement with the rigid rotor model breaks down at the $J = 8^+$
state for the B(E2)-values. Thus, when the $J = 0^+$ and $J = 2^+$
two-particle states are not so dominant a factor, the rotational
pattern is violated.

A similar analysis of the $K = 0_1^+$ band of the 4 x 4 system
leads to similar conclusions. There, the "rotational behavior"
and the dominance of the $J = 0^+$ and $J = 2^+$ states deteriorates
after the $J = 4^+$ state in the 4 x 4 system. These results suggest
that when additional two particles are added, the collective behavior extends to two higher units in angular momentum.

The $K = 0_2^+$ band in the (4 x 6) system is dominated by $J = 0$
and $J = 2$ couplings up to the $J = 4^+$ state. The states in the
$K = 2$ band in both the (4 x 4) and (4 x 6) systems are much more
complicated than are the states in the $K = 0_1^+$ band. In the $K = 2$
bands, the states are not dominated by the $J = 0$ and $J = 2$
couplings. In particular, there are significant admixtures of
the lowest $J = 3^+$ two-particle eigenstates.

In summary, the calculations reported here offer considerable
supporting evidence for the validity of the assumptions of the
Interacting Boson Model. The calculations show:

1. The coupling of key low-lying neutron and proton states
of valence particles can lead to collective rotational features
which appear to be more distinct and which extend to higher
angular momenta as the number of particle increases.

2. Those states which are rotational are dominated by states
formed by coupling $J = 0$ and $J = 2$ two-particle states.

References

1. K. T. Hecht, J. B. McGrory, and J. P. Draayer, Nucl. Phys.
   A197, 369 (1972).
2. I. M. Green and S. A. Moszkowski, Phys. Rev. 139B, 790 (1965).

ELECTRON SCATTERING IN THE INTERACTING BOSON MODEL

A.E.L. Dieperink

Kernfysisch Versneller Instituut

Groningen, the Netherlands

1. INTRODUCTION

From the preceeding talks two different aspects of the interacting boson model (IBA) of collective degrees of freedom in nuclei
have emerged. First the symmetry aspects of the model have been
emphasized[1]. It was shown that the Hamiltonian can be expressed
in terms of the generators of the SU(6) group, and in special
cases in terms of subgroups of SU(6); the application of group
theoretical methods then provides an elegant and simple description
of energy levels. Similarly the calculation of electromagnetic
transition rates becomes straightforward. For example, it was
shown that the one-boson E2 operator (neglecting separate neutron-
proton degrees of freedom)

$$\hat{T}_\mu^{(2)} = \alpha_2 (s^+ d_\mu + d_\mu^+ s)^{(2)} + \beta_2 (d^+ d)_\mu^{(2)}, \tag{1}$$

with constants α_2 and β_2 could describe experimental B(E2) values
and quadrupole moments rather well.

On the other hand, it is well known that the operator (1)
merely constitutes a special case, namely the long wave-length
limit, of a more general quadrupole transition operator

$$\hat{T}_\mu^{(2)} \underset{q \to 0}{\sim} \lim q^{-2} \int dr \, r^2 \, j_2(qr) \hat{\rho}_\mu^{(2)}(r), \tag{2}$$

where the multipole charge density operator $\hat{\rho}^{(\lambda)}(r)$ has been introduced:

$$\hat{\rho}_{\mu}^{(\lambda)}(r) = \int d\Omega \; \hat{\rho}(\vec{r}) Y_{\lambda\mu}(\Omega) \tag{3}$$

The operators $\hat{\rho}^{(\lambda)}(r)$ enter in the description of excitation processes of a nucleus via a one-body mechanism in which momentum q is transferred. Naturally the question arises whether these processes, in which the radial extension of the nucleus is probed in more detail, can also be described in the IBA. This seems especially interesting in connection with the second aspect of the IBA that has been discussed[2]: the relation between the structure of the boson and the fermion pairs that make up the boson. Such a microscopic foundation of the IBA indeed suggests that the matrix elements of the more general operator (3) may be treated by the IBA. This would open the interesting possibility to describe excitation functions of low lying collective states in nuclei in a simple, yet fundamental, manner.

In this talk I will first explore the possibility to describe quadrupole transition densities in the IBA at a phenomenological level (section 2), a similar approach to the matrix elements of the E0 operator is discussed in section 3. In section 4 I will briefly outline a possible microscopic approach to obtain the radial structure of the boson densities, and conclude with some general remarks.

2. PHENOMENOLOGY OF E2 FORM FACTORS IN IBA

The electron scattering cross section can be expressed (in PWIA) in terms of the square of the form factor

$$F_{\lambda}(q) = \int dr \; r^2 \; j_{\lambda}(qr) < \psi_f || \hat{\rho}^{(\lambda)}(r) || \psi_i >, \tag{4}$$

where the multipole transition densities $< \psi_f || \hat{\rho}^{(\lambda)}(r) || \psi_i >$ contain the nuclear structure information. Cross sections for other direct reactions can also be expressed in these quantities although in general a more complicated weighting function is needed.

A straightforward generalization of eq.(1) to the description of transition densities leads to a one-body operator of the form [3]

$$\hat{\rho}^{(2)}(r) = \alpha_2(r)(d^+s + s^+d)^{(2)} + \beta_2(r)(d^+d)^{(2)} \tag{5}$$

The reduced matrix elements of $\hat{\rho}^{(2)}(r)$ between initial and final states can be written as

$$< [N]\chi'L' ||\hat{\rho}^{(2)}(r)|| [N]\chi L > = \alpha_2(r)A + \beta_2(r)B \equiv \rho^{(2)}(r)$$

(6)

where the boson densities are given by

$$A = < [N]\chi'L' ||d^+s + s^+d|| [N]\chi L >$$ (6a)

$$B = < [N]\chi'L' ||(d^+d)^{(2)}|| [N]\chi L >$$ (6b)

The assumption (5) has several implications, that can be tested experimentally. First, since every E2 transition density can be expressed in terms of only <u>two</u> functions [$\alpha_2(r)$ and $\beta_2(r)$ in eq.(5)] there should exist linear relations between collective transition densities in one nucleus. In particular, the transition densities for three $0 \to 2_i^+$ (i=1,2,3) transitions are related through

$$< 2_3^+||\hat{\rho}^{(2)}(r)||0^+ > = N_1 < 2_1^+||\hat{\rho}^{(2)}(r)||0^+ > +$$

$$+ N_2 < 2_2^+||\hat{\rho}^{(2)}(r)||0^+ >$$

(7)

(with arbitrary constants N_1, N_2).
Secondly, if in addition the structure of the wave functions of the states involved is known, i.e.

$$| [N]\chi LM > = \sum_{n_d\chi'} a(L\chi,n_d\chi')(d^+)^{n_d}_{L\chi'}(s^+)^{N-n_d}|0 >$$

(8)

the values of A and B in eq.(6) can be calculated, leading to a prediction for the values of N_1 and N_2.

Finally if the functions $\alpha_2(r)$, $\beta_2(r)$ do not vary from isotope to isotope, i.e. if shell effects are negligible, the variation in transition densities $< 2_i^+||\hat{\rho}^{(2)}(r)||0^+ >$ can be predicted, since in general the relative contributions of the d-boson changing and the d-boson conserving terms in eq.(5) will vary with neutron number. Although high quality experimental information from electron scattering is still rather limited it is tempting to apply these ideas to some existing data in the Sm-Nd region.

2.1 THE NUCLEUS ^{150}Nd

Recently at Bates inelastic excitation of states at 130, 850 and 1060 keV has been observed[4,5], which were known in the literature as states with spin 2^+ [6]. The nucleus ^{150}Nd is an example of a

transitional SU_5 – SU_3 nucleus[7]. The energy levels can be fitted
quite well in the frame work of IBA with a boson Hamiltonian
with four parameters: ε_d, P.P (pairing), Q.Q (quadrupole) and
H.H(hexadecapole) forces. The calculated level scheme is compared
with experiment in fig.1. The experimentally known $M(E2)$ matrix
elements, such as $< 2_1^+||M(E2)||0^+>$ and $<4_1^+||M(E2)||2_1^+ >$ can be
reproduced rather well with the one-body quadrupole operator of
eq. (1) with values of α_2 = 14.3 efm^2, β_2 = -.67 α_2. The B(E2↑)
values for the $0^+ \to 2_1^+$ transitions are given in table 1. Note that
the calculated $B(E2,0\to2_2^+)< B(E2,0\to2_3^+)$ in agreement with experiment.

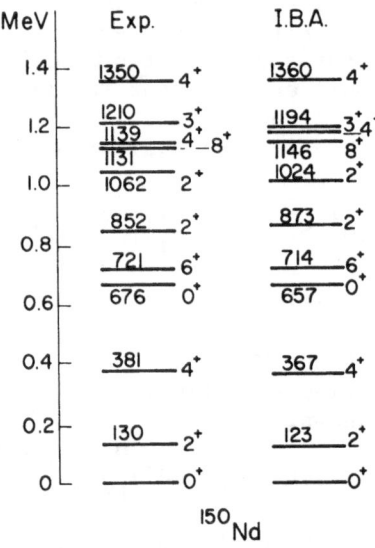

Fig.1.   Low lying levels in ^{150}Nd
         compared with the result of
         the IBA, calculated with
         the program PHINT.

    The electron scattering data from Bates were analysed in DWBA.
For convenience the transition densities were parametrized in the
form $\rho^{(2)}(r) = \sum_n a_n r^n e^{-\gamma r^2}$. Cross sections taken at a fixed angle
(90°) but for various electron energies have been transformed to a
common energy E_0 (the highest measured one), and are calculated as
a function of the effective momentum transfer

$$q_{eff} = q(1 + \frac{3Z\alpha\hbar c}{2E_0 R_0})$$

Table 1. One-body boson densities A_i and B_i, and B(E2) values for the $0 \rightarrow 2_i^+$ transitions in ^{150}Nd.

State	A_i	$B_i/\sqrt{5}$	$B(E2\uparrow)_{calc}$	$B(E2\uparrow)_{exp}[e^2fm^4]$
2_1^+	10.02	−1.01	27196	27200
2_2^+	1.68	0.58	134	76
2_3^+	2.73	0.65	629	690

In fig.2 the DWBA fit to the experimental form factor F(q) as $F(q) = [\frac{d\sigma}{d\Omega} / (\frac{d\sigma}{d\Omega})_{Mott}]^{1/2}$ is shown for the two stronger states, 2_1^+ and 2_3^+.

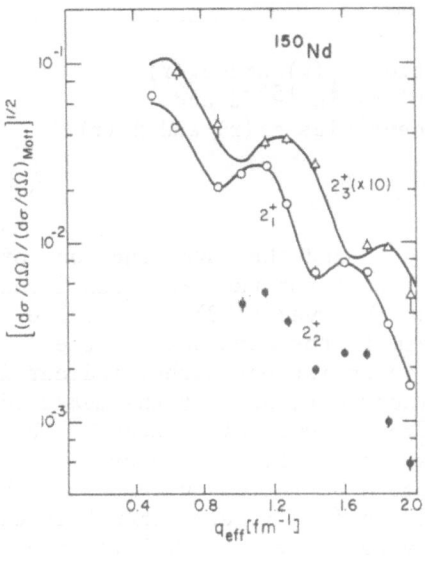

Fig.2 The experimental form factors $F(q_{eff}) = (\sigma/\sigma_{Mott})^{\frac{1}{2}}$ for the inelastic excitations of the lowest three 2^+ states in ^{150}Nd. The full lines show the fits to the $F_{2_1^+}$ and $F_{2_3^+}$.

The extracted E2 transition densities and the corresponding boson densities $\alpha_2(r)$ and $\beta_2(r)$ are shown in figure 3. The d-boson changing function $\alpha_2(\bar{r})$ has the shape characteristic for a collective transition density peaked at the nuclear surface, whereas $\beta_2(r)$ has a more complicated structure, and a significantly larger transition radius R_t, where

$$R_t = \int \rho^{(2)}(r)r^6dr / \int \rho^{(2)}(r)r^4dr.$$

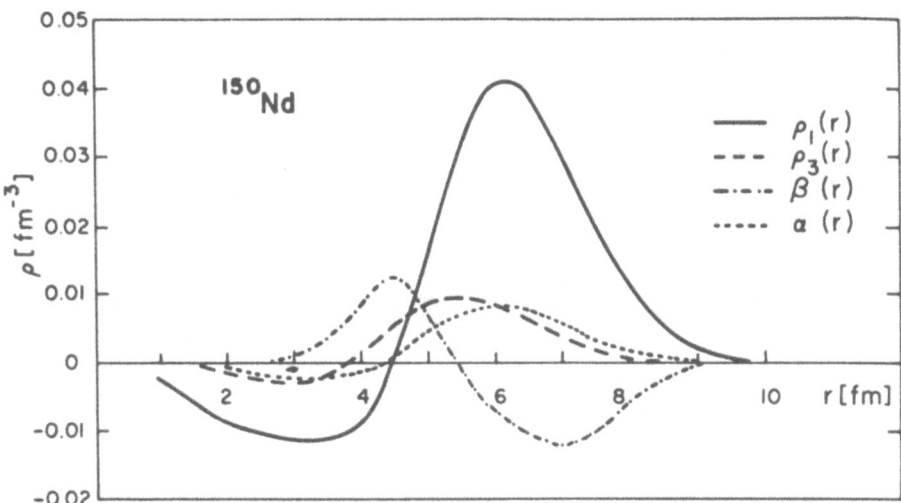

Fig.3. The transition densities $\rho_1(r)$ and $\rho_3(r)$
for the $0_1^+ \to 2_3^+$ transitions in ^{150}Nd and
the extracted boson densities $\alpha_2(r)$ and $\beta_2(r)$.

Using the extracted $\alpha_2(r)$ and $\beta_2(r)$ and the wave functions of
^{150}Nd one can then predict the form factor for the $0 \to 2_2^+$ transition.
It is seen in fig.4 that although the observed B(E2) value is reproduced the predicted quadrupole form factor does not fit the
reported cross section at all. This inconsistency either indicates
that the parametrization (eq.5) is incorrect, or that the measured
form factor contains contributions from an unresolved state. The
latter hypothesis is supported by two observations: the cross
section for the $E_x = 850$ keV state could be fitted quite well with
an L=1 transition density and also in an (n,n'γ) experiment[8] it was
found that the 850 keV state actually consists of a $(1^-, 2^+)$ doublet.
From a more detailed analysis[4] it appears that a consistent description can be obtained for the excitation of the 1^- state and the
3^- state at 931 keV in terms of an octupole $K^\pi = 0^-$ band.
In passing it seems worth to point out that the rather strong E1
form factor at higher q is not inconsistent with the reported
small B(E1) value ($\sim 10^{-3}$W.u.). The explanation is that the isoscalar
contribution to the operator ($\sim \sum j_i (qr_i)$) should vanish in the limit
$q \to 0$ because of translational invariance. Therefore at $q \sim 0$ the contribution to the E1 matrix element should be isovector in nature,
and therefore small, whereas at higher q this constraint is not
effective.

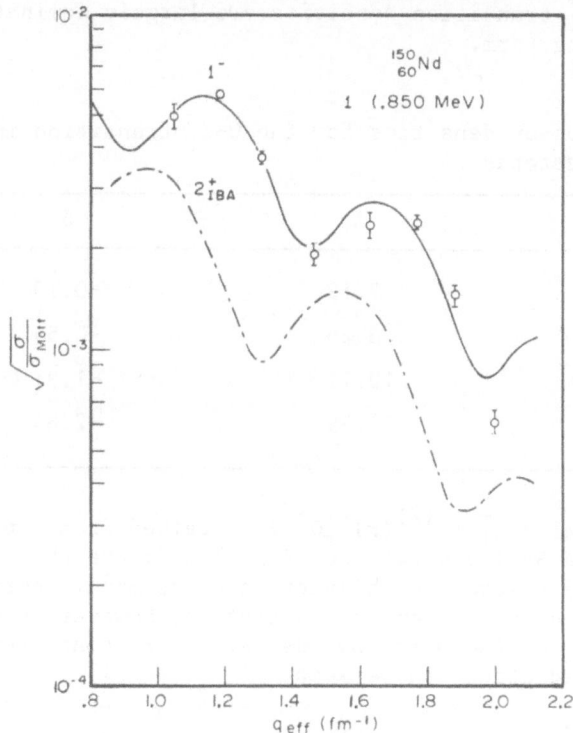

Fig.4. Electron scattering form factor for the 850 keV state in ^{150}Nd. The solid curve results from a calculation in which the state is represented as a $J^{\pi}(K) = 1^{-}(0)$ octupole surface oscillation belonging to the same band as the $3^{-}(931$ keV) state. The dot-dashed curve represents the IBA 2^{+}_{2} prediction.

2.2 Sm ISOTOPES

As mentioned above a weaker test of the form of the operator (5) for electron scattering consists of examining the trend in $0 \rightarrow 2^{+}_{1}$ transitions in a series of isotopes. Recently these transitions have been measured for several Sm isotopes by the Tél Aviv-Saclay group[10]. It was shown by Scholten[7,9] that the trend in the B(E2, $0 \rightarrow 2^{+}_{1}$) values in the Sm isotopes could be described quite well in terms of the operator (1). In table 2 the values of the one-body boson densities $A_1 = < 2^{+}_{1} || d^{+}s + s^{+}d || 0^{+} >$ and $B_1 = < 2^{+}_{1} || (d^{+}d)^{(2)} || 0^{+} >$ are given for the Sm isotopes[7]. It is seen that as a function of N the A_1 values increase almost linearly whereas the B_1 values increase more quadratically with N. On the basis of the information about $\alpha_2(r)$. and $\beta_2(r)$ obtained from ^{150}Nd one thus expects an increase in the transition radius going to the heavier isotopes; such an effect would be rather small since all $0^+ \to 2_1^+$ transition densities are largely dominated by the d-boson changing term.

Table 2: One-boson densities for the $0 \to 2_1^+$ transition in the Sm isotopes

Nucleus	A_1	B_1
^{148}Sm	7.12	-0.19
^{150}Sm	8.26	-0.55
^{152}Sm	10.71	-1.91
^{154}Sm	11.99	-2.89

The experimental $< 2_1^+ || \hat{\rho}^{(2)}(r) || 0^+ >$ obtained from a preliminary analysis of the Saclay data[10] (see fig.5) indicate that the increase in collectivity towards the heavier isotopes mainly corresponds to an overall shape independent normalization; however, a small shift in the position of the peak towards larger r consistent with the qualitative prediction can be seen.

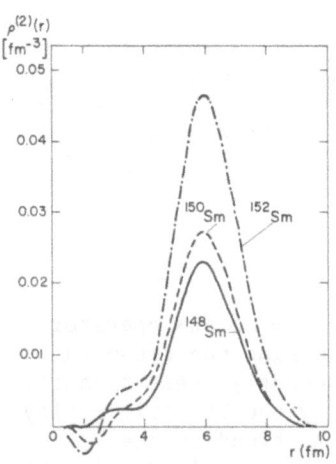

Fig.5. Transitions densities for the $0 \to 2_1^+$ transitions in ^{148,150,152}Sm.(Preliminary results[10]).

I conclude that although the available data do not allow for a quantitative comparison the IBA model provides a convenient parametrization of electric quadrupole form factors, and leads to predictions for cross sections.

We note that certain moments of the functions $\alpha_2(r)$ and $\beta_2(r)$ also enter in the analysis of muonic X-ray data. As an example we mention the results[15] for the equivalent quadrupole radii R_m^{22} and R_m^{20} for the $< 2_1^+||\hat{\rho}^{(2)}(r)||2_1^+>$ and $< 2_1^+||\hat{\rho}^{(2)}(r)||0^+>$ densities in ^{152}Sm. The extracted value[15] of $\delta R = -0.05 \pm 0.11$ fm is consistent with the almost vanishing value predicted by the IBA for a SU_3 type nucleus.

3. E0 CASE

The evaluation of the E0 operator in the long wave length limit, $< r^2 >$, in the IBA has been discussed by Arima et al.[11]. In lowest order the E0 transition density operator can be expressed as

$$\hat{\rho}^{(0)}(r) = \hat{\rho}_{core}^{(0)}(r) + \alpha_0(r)(s^+s)^{(0)} + \beta_0(r)(d^+d)^{(0)} \tag{9}$$

By using boson number conservation $N = n_s + n_d$ eq.(9) can be rewritten as

$$\hat{\rho}^{(0)}(r) = \hat{\rho}_{core}^{(0)}(r) + \alpha_0(r)\,\hat{N} + \beta_0'(r)(d^+d)^{(0)} \tag{10}$$

It then follows that E0 transitions and isomer shifts can be expressed in terms of the last term of eq.(10), whereas isotope and isotone shifts receive contributions from the last two terms. The values $c_N^L = < L||(d^+d)^{(0)}||L >$ as obtained in Groningen[14] from a fit to the energy spectra of the Sm isotopes are given in table 3 for the ground state and the first 2^+ state.

Table 3: Some monopole matrix elements for the Sm isotopes

Nucleus	c_N^0	c_N^2
^{148}Sm	.092	.57
^{150}Sm	.289	.79
^{152}Sm	1.38	1.56
^{154}Sm	2.22	2.27
^{156}Sm	2.30	2.33

As an illustration the result for the isomer shift are shown [7] in fig.6:

$$\delta < r^2 > = (c_N^2 - c_N^0)\beta_0' , \qquad (11)$$

where $\beta_0' = \frac{1}{Z} \int \beta_0(r) r^4 dr = .09 \ fm^2$.

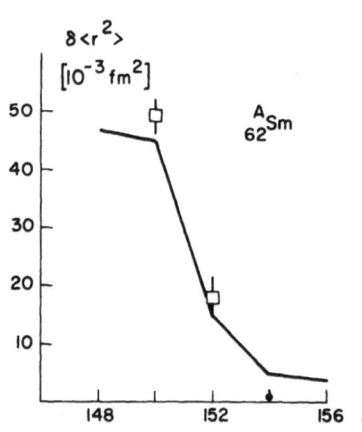

Fig.6. Comparison between calculated (full line) and experimental (points) isomer shifts in the Sm isotopes. The experimental points are taken from refs.(15) (A=150,152) and ref.(7)(A=154).

Note that the large drop in the isomer shift towards the SU_3 region is very well reproduced. The functions $\alpha_0(r)$ and $\beta_0'(r)$ can in principle be determined from elastic electron scattering on the various Sm isotopes. A preliminary analysis of the data of the TelAviv-Saclay group[10] in terms of eq.(10) is not conclusive.

The special case of the isotope shift

$$\Delta < r^2 > = < r^2 >_A - < r^2 >_{A-2} = \alpha_0 + \beta_0' \ (c_N^0 - c_{N-2}^0), \qquad (12)$$

with $\alpha_0 = \frac{1}{Z} \int \alpha_0(r) r^4 dr$, is compared with experiment in fig.7; for β_0' the same value as for the isomer shift has been used. Additional information on $\beta_0'(r)$ could come from measurements of monopole form factors $0_1^+ \rightarrow 0_2^+$ transitions.

It should be noted that the IBA approach to the description of monopole matrix elements is rather different from other methods. For example, in a microscopic approach to the calculation of isomer shifts[12] it was stressed that the net result depends on a delicate trade off of contributions of various single proton orbitals near the Fermi surface. On the other hand there are also indications, e.g. from the observed odd-even staggering of isotope shifts[13] that correlated pairs of nucleons - the building blocks of IBA - play an important role in monopole properties.

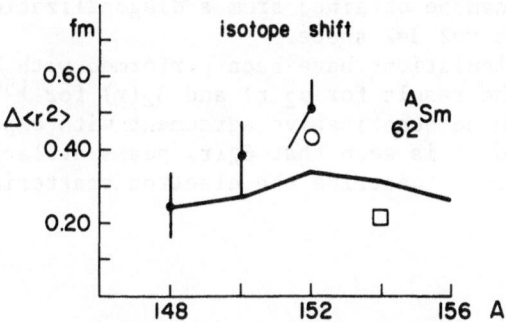

Fig.7 Comparison between calculated (full line) and
experimental isotope shifts for the Sm isotopes.
The experimental data are from ref.(16)(points),
ref.(15)(open circle) and ref.(22)(cross).

4. MICROSCOPIC ASPECTS

An important goal of the present approach, of course, is to
construct the functions $\alpha_\lambda(r)$, $\beta_\lambda(r)$ from a microscopic point of
view following the methods outlined by Iachello[1] . Recently, Otsuka
et al.[17] have given a general derivation of the boson Hamiltonian
(and other operators) starting from a fermion Hamiltonian (and
operators). The effect of many quasi-degenerate j shells was
simulated by a single large j shell. To describe the radial depen-
dence of $\alpha_\lambda(r)$ and $\beta_\lambda(r)$ one needs to consider wave functions with
different and realistic radial behaviour[18] .
For given pair operators

$$S^+ = \sum_j \alpha_j \sqrt{\frac{2j+1}{2}} [a_j^+ a_j^+]^{(0)} \tag{13}$$

$$\text{and} \quad D^+ = \sum_{ij} \beta_{ij} [a_i^+ a_j^+]^{(2)} \tag{14}$$

the quadrupole transition densities are given by

$$r^4 \alpha_2(r) = < S||\sum_i r_i^2 \delta(r-r_i) Y_2(\hat{r})||D > , \tag{15}$$

$$\text{and} \quad r^4 \beta_2(r) = < D||\sum_i r_i^2 \delta(r-r_i) Y_2(\hat{r})||D > . \tag{16}$$

For the Sm-Nd region the relevant proton orbitals are the $g_{7/2}$,
$d_{5/2}$, $d_{3/2}$, $s_{1/2}$ and $h_{11/2}$ orbitals. The α_i can be determined by
diagonalizing the pairing Hamiltonian in the v=0 space. For the construction of the D^+ operator several prescriptions can be used;[18] e.g.

one may define $D^+ \equiv [Q,S^+]$ where Q is the quadrupole operator, or the coefficients β_{ij} can be obtained from a diagonalization of the Hamiltonian in the $v=2$ $J=2$ space.
Some schematic calculations have been performed with harmonic radial wave functions. The result for $\alpha_2(r)$ and $\beta_2(r)$ for ^{150}Nd are shown in fig.8. Although no quantitative agreement with experiment (fig. 3) can be expected it is seen that $\beta_2(r)$ peaks at larger r than $\alpha_2(r)$, as is needed to describe the electron scattering.

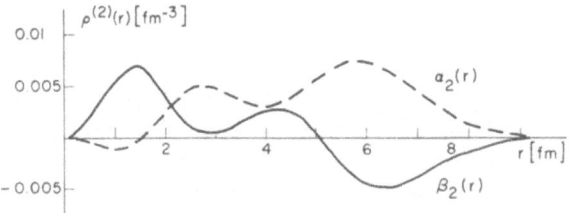

Fig.8. Quadrupole transition densities $\alpha_2(r)$ and $\beta_2(r)$ calculated for the Nd-Sm region

5. DISCUSSION

Other attempts to describe form factors in terms of collective coordinates have been made by Borie et al.[19] in terms of the collective Hamiltonian of Greiner et al.[20]. In ref.(20) the collective wave functions are obtained from a diagonalization of a collective "adiabatic" Hamiltonian in the collective coordinates $\alpha_{2\mu}$

$$H \simeq B\dot{\alpha}_2\dot{\alpha}_2 + V(\alpha_2^2, \alpha_2^3, \alpha_2^4)$$

Subsequently the transition density operator is obtained by a Taylor series expansion of the density $\hat{\rho}(\vec{r}-\vec{R})$, with radius $\vec{R} = R_0(1 + \alpha_{20}Y_{20})$ in terms of α_2. Apart from the observation that a connection to a microscopic picture is difficult to make it is clear that this approach differs from the IBA one in several respects, e.g. (i) the Hamiltonian in ref.(19) is restricted to terms quadratic in the kinetic energy, (ii) to describe collective E_2 transitions the calculation of matrix elements of operators of high order in α_2 like $(\alpha_2 \hat{} \alpha_2)^{(2)}, ((\alpha_2 \backsim \alpha_2) \backsim \alpha_2)^{(2)}$, etc. is required. It can be shown that part the effects of these higher-order terms via the introduction of the s-boson in IBA is absorbed into the one-body boson operator $(d^+s + s^+d)$; on the other hand terms of the type d^+d^+ss which are effectively present in the conventional approach, are absent in the lowest order IBA.

It is clear that electron scattering is a powerful tool to investigate the validity of the IBA approach. Some selective tests of the model of the type discussed above would be very useful.

Finally I want to point out that the introduction of separate degrees of freedom for neutron and proton bosons[21] leads to new interesting speculations. The generalized IBA model predicts low-ying collective states which are not symmetric in neutron and proton components. In other words for these states one would expect different transition matter and charge distributions. Experimentally these could be distinguished by comparing nuclear reactions with probes that interact in different ways with neutrons and protons (electrons, alpha particles, $\pi^{\pm}$ etc.). Also the generalized IBA predicts low-lying collective 1^{+} states not present in the original formulation. In electron scattering such states should show up as M1 transitions characterized by a strong convection current and a small magnetization current contributions.

References

1. F.Iachello, contribution to this meeting
2. T.Otsuka, contribution to this meeting
3. A.E.L.Dieperink et al., Phys.Lett.76B(1978)135
4. C.Creswell et al., Phys.Rev.C18(1978)2432
5. C.Creswell et al., to be published
6. Nuclear Data Sheets 18(1976)233
7. O.Scholten, A.Arima and F.Iachello, Ann.Phys.(N.Y.)115 (1978)325
8. D.F.Cooper et al., Progress Report(1974-1977),Univ.of Kentucky, p.70
9. O.Scholten, contribution to this meeting
10. N.Haig et al., to be published and M.Moinester,private communication
11. A.Arima and F.Iachello,Ann.Phys.(N.Y.) 99(1976)253;111(1978)201
12. J.Meyer and J.Speth, Nucl.Phys.A203(1973)17
13. G.Nowicki, Phys.Rev.C18(1978)2369
14. F.de Jong and F.Iachello, private communication
15. Y.Yamazaki et al., Phys.Rev.C18(1978)1474
16. R.J.Powers et al., preprint CALT-63-297
17. T.Otsuka, A.Arima and F.Iachello,Nucl.Phys.A309(1978)1
18. T.Otsuka, private communication
19. E.Borie, D.Drechsel and K.Lezuo, Nucl.Phys.A211(1973)393
20. G.Gneuss and W.Greiner, Nucl.Phys.A171(1971)449
21. T.Otsuka et al., Phys.Lett.76B(1978)139
22. Nuclear Data Sheets 14(1974)613

NUCLEAR FIELD THEORY TREATMENT OF THE INTERACTING BOSON MODEL

D. R. Bes* and R. A. Broglia**

State University of New York
Department of Physics
Stony Brook, New York 11794

ABSTRACT

A microscopic description of the interacting boson model is attempted in a basis of multipole pairing vibrations and in the framework of the nuclear field theory. If nothing else, it seems in this way possible to calculate both rotations and vibrations in a common basis.

Both fermionic and bosonic excitations play an important role in determining the properties of the nuclear spectrum[1]. A mathematically rigorous treatment of the interweaving of these elementary modes of excitation has been provided by the nuclear field theory[2] (NFT), for systems with no broken symmetries.

It has been recently recognized[3] that, in many cases of interest such as two phonon systems, already the lowest order Tamm-Dancoff (TD) NFT-contributions give a rather accurate approximation to the exact solution when calculated in the Rayleigh-Schrödinger perturbation theory. This result has advanced the possibilities of carrying out realistic calculations in systems with open shells of both protons and neutrons. Even so, to make the problem of soft and deformed nuclei tractable it is necessary to resort to models. One such model

---

* Permanent address: Comisión Nacional de Energia Atómica, Buenos Aires, Argentina. Fellow of the Consejo Nacional de Investigaciones Cientificas y Tecnicas.
**Permanent address: Niels Bohr Institute, University of Copenhagen, Denmark.

is the interacting boson model[4] (IBM), which provides a rather successful systematization of a number of nuclear properties. It is
noted that, although main features of the spectra are reproduced, it
is still not understood why. The results suggest anyhow that there
is something clever in the way the model takes into account the
pairing and quadrupole degrees of freedom.

In the IBM model the degrees of freedom of particles outside
closed shells are described in terms of four bosons, two for protons
and two for neutrons, carrying angular momentum and parity 0^+ (s-
bosons) and 2^+ (d-bosons). The only interaction between the bosons
is

$$H_{\pi\nu} = -\kappa_{\pi\nu}\sqrt{5}(Q_2(\pi)\ Q_2(\nu))_0 ,$$ (1)

where $Q_{2\mu}$ is the quadrupole operator.

The phonons of the IBM have been interpreted[5] as pairs of like
particles coupled to J=0 and J=2. Making use of a fermion mapping,
which seems to be unique at least for a single j-shell, the matrix
elements of (1) are corrected by introducing external Pauli principle conserving factors. The same correction is applied to the different physical operators. Note, however, that in most of the
calculations (cf. e.g. ref. 6), where the so-called phenomenological model was utilized, no such corrections have been taken into
account. It can be argued that the actual values of the parameters
take into account the Pauli principle in an average way.

The model Hamiltonian has two parameters, i.e. the energy difference between the s and the d bosons, and the strength $\kappa_{\pi\nu}$ of the
proton-neutron interaction (1).

In this letter we give a microscopic interpretation of the IBM.
In particular Pauli principle corrections and overcompleteness of the
basis are systematically treated.

We consider a system of multipole proton and neutron pairing
vibrations (cf. e.g. refs. 7 and 8). The properties that completely
determine the pairing boson fields are the particle-vibration coupling strength Λ (cf. fig. 1(A)) and the energy ω. For a j-shell and
a separable pairing force we obtain, in the TD approximation, (cf.
eqs. (3.1) - (3.5) ref. 2)

$$\Lambda_\lambda \equiv \Lambda(2;\lambda) = -(4\pi)^{\frac{1}{2}} G_\lambda\ \langle j||T_\lambda||j\rangle ,$$ (2)

and

$$\omega_\lambda \equiv \omega(2;\lambda) = \varepsilon - (2\pi)^{\frac{1}{2}} G_\lambda\ \langle j||T_\lambda||j\rangle ,$$ (3)

where $T_{\lambda\mu} = f_\lambda(r)\ Y_{\lambda\mu}(\hat{r})$, $f_\lambda(r)$ being a smooth function of r peaked

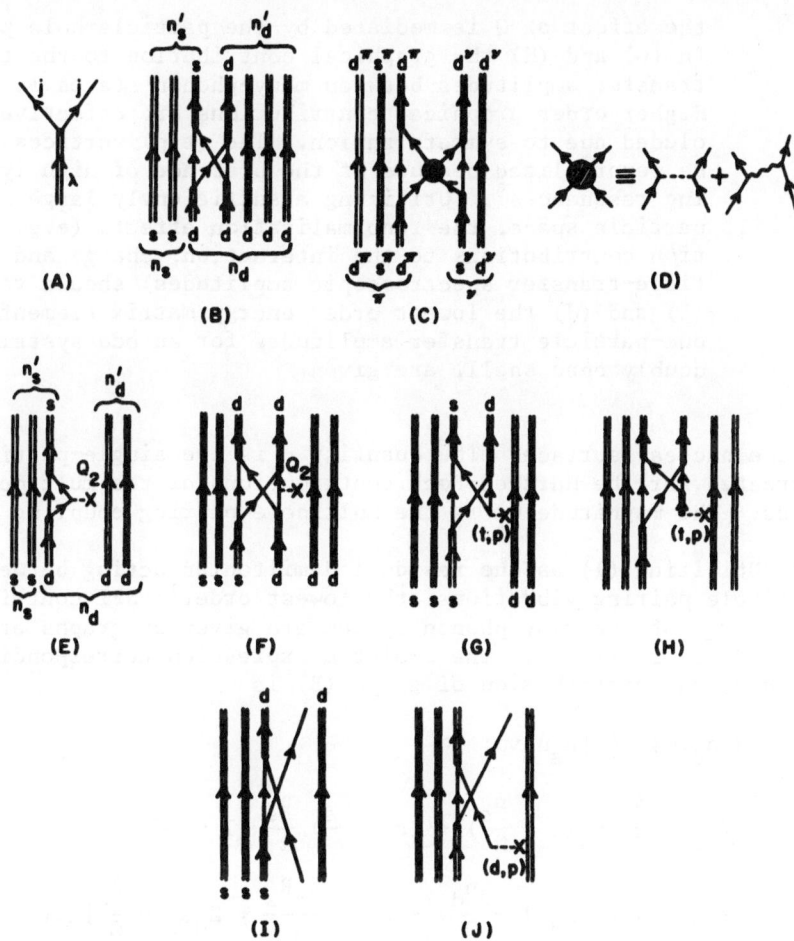

Fig. 1. Nuclear field theory diagrams describing the interaction and
transfer amplitudes of a system of monopole (s) and quadrupole (d) pair addition modes. In (A) the coupling between
the pairing mode of multipolarity λ and the fermions moving
in a j-orbital is displayed. The pairing phonons are drawn
as a double arrowed line while a single arrowed line represents a fermion. The Pauli principle correction between
like phonons (i.e. proton-like or neutron-like phonons) is
shown in (B). The many-phonon character of the interaction
arises from the symmetrization of the resulting expression
between particles and holes (cf. text). In (C) the quadrupole effective interaction between proton-like and neutron-
like phonons is shown. The contributions to it arising from
the bare model interaction and the exchange of a quadrupole
(particle-hole) phonon are given in (D). In (E) and (F) are
collected the quadrupole transition amplitudes between many
phonon states. Effective charges arise, as in (D), because

the effect of Q is mediated by the particle-hole phonons. In (G) and (H) the graphical contribution to the two-nucleon transfer amplitudes between many-phonon states is shown. Higher order graphical contributions are effectively included due to symmetrization. The (t,p) vertices can also be renormalized because of the presence of high lying pairing resonances[9]. Utilizing a sufficiently large single-particle space, the renormalization effects (e.g. polarization contributions to the interaction, charge and two particle-transfer spectroscopic amplitudes) should vanish. In (I) and (J) the lowest order energy matrix elements and the one-particle transfer amplitudes for an odd system with doubly open shell, are given.

at the nuclear surface. The quantity ϵ is the single-particle energy corrected for the Hartree-Fock contributions of the multipole pairing force. The magnitude G_λ is the multipole pairing coupling constant[10].

Utilizing (1) as the residual Hamiltonian acting between the multipole pairing vibrations, the lowest order[11] NFT contributions to the energy of the many phonon system are given by graphs of the type (B) and (C) in fig. 1. The analytic expression corresponding to e.g. the diagonal contribution of graph (B) is

$$\langle n_s n_d v\alpha;\ I|H|n_s n_d v\alpha;\ I\rangle$$

$$= (\omega_s - \frac{Z_s}{\Omega})n_s(1 - \frac{n_s}{\Omega}) + (\epsilon - \frac{Z_s}{\Omega})\frac{n_s}{\Omega}(1 - \frac{n_s}{\Omega})^2$$

$$+ (\omega_d - Z_d R_2)n_d(1 - \frac{n_d}{\Omega}) + [\frac{\epsilon}{\Omega} - \frac{Z_d R_2}{\Omega} + Z_d R_2 - \frac{Z_d}{\Omega}]n_d^2(1 - \frac{n_d}{\Omega})^2$$

$$+ [n_s(1 - \frac{n_s}{\Omega})n_d(1 - \frac{n_d}{\Omega})]\ (Z_s + Z_d)\ 2/\Omega\ ,\qquad (4)$$

where

$$R_2 = 50 \sum_{v''\alpha''I''I'''} (n_d-2\ v''\alpha'',I'';I'''|\} n_d\ v'\alpha';I)$$

$$(n_d-2\ v''\alpha'',I'';2|\}n_d\ v\alpha;I) \begin{Bmatrix} j & j & 2 \\ j & j & 2 \\ 2 & 2 & I'' \end{Bmatrix}.\qquad (5)$$

The many-phonon basis states are labeled by the numbers n_s and n_d of s and d pairing vibrations, by the boson seniority v, by the angular momentum I and by an extra label α needed because of the degeneracies of the many phonon spectrum (cf. ref. 12). The quantities appearing in eq. (5) are, aside from the 9-j symbol, two-boson fractional parentage coefficients[12]. The particle-vibration coupling strength (2) and the correlation energy of the phonons enter in the matrix elements through the quantity $Z_\lambda = \varepsilon - \omega_\lambda = \Lambda_\lambda <j||T_\lambda||j>/\sqrt{2}$. The expression (4) has been symmetrized with respect to the interchange of particles and holes. This is also done for non-diagonal contributions.

For the particular case of s-bosons, the sum $<n_s|H|n_s>$ of the zeroth-plus the first-order contributions of the diagrammatic expansion in $1/\Omega$, already gives the exact answer[13]. In this case, the phonon energy is $\omega_s = G_o(1-\Omega)$, while graph (B) of fig. 1 is equal, for the case of two s-bosons, to $2G_o$. Thus,

$$<n_s|H|n_s> = n_s\omega_s + 2G_o[\frac{n_s(n_s-1)}{2}] = -G_o n_s(1 - \frac{n_s}{\Omega})\Omega. \tag{6}$$

A basic feature of this expression is that it is symmetric with respect to the interchange of particles and holes.

In a more general case, this symmetry, would arise from the contributions mentioned above, as well as from higher order contributions. The resulting expression to order n_λ^2 can be written as

$$A\, n_\lambda(1 - \frac{n_\lambda}{\Omega}) + A\, \frac{n_\lambda^2}{\Omega} + B\, n_\lambda^2.$$

The coefficient B corresponds to the graphical contributions of order $1/\Omega$ with respect to the leading term. Thus A/Ω and B are of the same order. Moreover, they tend to cancel out. In particular, this cancellation is exact in the case of s-bosons. The same reasoning can be applied to the higher order Pauli corrections.

A simplified, albeit approximate, way to obtain the symmetrized expressions is to replace in each case the leading contribution $A\, n_\lambda$ by $A\, n_\lambda(1 - n_\lambda/\Omega)$, the next order contribution $B\, n_\lambda^2$ by $(B + A/\Omega)n_\lambda^2(1 - n_\lambda/\Omega)^2$, etc.

The lack of orthogonality of the many-phonon states, as well as the Pauli principle violations are consistently treated and on equal footing in the NFT through e.g. graphs of type (B) of fig. 1.

The energy matrix elements between proton and neutron pairing vibrations (cf. eq. (1)) are given by graph (C) of fig. 1. The corresponding analytic expressions, symmetrized between particles and holes, can be written in terms of the matrix elements (E) and (F) (cf. fig. 1) of the quadrupole operator[14] $Q_{2\mu} = \Sigma<j||Q_2||j> [a^+_j a_j]_{2\mu}$. The analytic expression corresponding to e.g. the non-diagonal contribution of graph 1(E), is

$$<n_s+1\ n_d-1\ v'\alpha';I'||Q_2||n_s\ n_d\ v\alpha;I>$$

$$= (n_s-1)^{\frac{1}{2}}(1 - \frac{n_s-1}{\Omega})^{\frac{1}{2}}(n_d)^{\frac{1}{2}}(1 - \frac{n_d}{\Omega})^{\frac{1}{2}} S_1 , \tag{7}$$

where

$$S_1 = 4\left(\frac{2I+1}{5}\right)^{\frac{1}{2}} (-1)^{I+I'} (n_d-1 \; v'\alpha';I'||\}n_d \; v\alpha;I)$$

$$\times \; <j||Q_2||j> \; , \tag{8}$$

A similar expression is obtained for $<n_s n_d v'\alpha'; \; I'||Q_2||n_s n_d v\alpha;I>$.
The quantity $<j||Q_2||j>$ is the single-particle quadrupole reduced
matrix element. Both the energy matrix elements between like phonons
as between protons and neutrons can change the number of d-bosons by
±2, ±1, and 0. The matrix element of (1) is thus written in terms of
matrix elements of $Q_{2\mu}$ like (7). It relates the static quadrupole
moment, the E2-transition amplitude as well as the effective charge
of the pairing modes to the effective interaction between protons and
neutrons (cf. refs. 1 and 16). The quantity $\kappa_{\pi\nu}$ is an effective
quadrupole coupling constant, as shown in graph (D) of fig. 1 (cf.
again ref. 16).

Diagonalizing the energy matrix with the normalization condition[17]

$$\sum_{ij} \{X_i X_j [\delta(i,j) - \frac{\partial H_{ij}}{\partial E}]\} = 1 \; , \tag{9}$$

where X_i is the amplitude of the basis state $|i> = |n_d n_s \; v\alpha;I>$ in the
final, correlated state, one obtains solutions which satisfy the
Pauli principle.

Once the amplitudes X_i are known, the matrix elements of the
quadrupole operator $Q_{2\mu}$ between initial and final states can be calculated in terms of the graphs of type (E) and (F) of fig. 1. For
the case of the two-particle-transfer operator $T_{\lambda\mu} = \Sigma_j [a^+_j a^+_j]_{\lambda\mu}/\sqrt{2}$,
the corrections to the leading order transfer amplitudes can be
worked out in terms of graphs (G) and (H) of fig. 1.

Note the two sources of Pauli principle corrections for the inelastic scattering and two-particle transfer amplitudes. The first
is due to the energy graphs (B), i.e. Pauli principle corrections in
the many-phonon basis states. The second is due to graphs (F) for
$<f||Q_2||i>$ and to both (G) and (H) for $<f||T_2||i>$. Both types of
corrections are expected to be of the same order of magnitude. In
fact, the importance of the second effect has been recently checked[3].

Finally, odd nuclei can be incorporated within the NFT framework discussed above. The energy matrix elements and the spectroscopic amplitudes associated with the one-particle transfer processes
are given by the many-phonon graphs (I) and (J) of fig. 1. Both are
Pauli principle corrections and have been utilized in e.g. ref. 18.

Because of the many pairing phonons which are utilized to describe doubly open shell nuclei, the quadrupole matrix elements which generate the deformed field[1] are in principle taken into account, although particle-hole excitations are apparently not utilized. In fact, they determine the strength of the effective interaction $\kappa_{\pi\nu}$. To the extent that the s boson can be interpreted as a monopole pairing mode which is specifically excited in two nucleon transfer processes, and which gives rise to a pairing condensate in single open shell nuclei, the competition between pairing and quadrupole fields (cf. ref. 1) should also be present in the model. Thus graph (I) of fig. 1 should in principle lead, for many j-shells and for large values of n^{π} and n^{ν} to the Nilsson scheme of levels[19].

A microscopic unified description of the nuclear spectrum, at least for calculation purposes, has been attempted by treating on equal footing and consistently s and d pairing bosons as well as fermions modes. Both nuclei with few particles outside closed shell as well as single and doubly open shell nuclei can be dealt on par, to lowest order in the nuclear field theory expansion parameter $1/\Omega$. The generalization of the model to many j-shells and the inclusion of non-adiabatic phonons, as well as excitations of other multipolarities appear to be within the reach of the present techniques.

Discussions with F. Iachello and O. Scholten are gratefully acknowledged.

REFERENCES

1. Cf. e.g. A. Bohr and B. R. Mottelson, Nuclear Structure, Vols. I and II, Benjamin (1969) and (1975).
2. Cf. e.g. P. F. Bortignon, R. A. Broglia, D. R. Bes, and R. Liotta, Phys. Rep. 30C:305 (1977), and references therein.
3. P. F. Bortignon, R. A. Broglia, and D.R. Bes, Phys. Lett. 76B:153 (1978).
4. A. Arima and F. Iachello, Phys. Rev. Lett. 40:385 (1978), and references therein.
5. A. Arima, T. Otsuka, F. Iachello, and I. Talmi, Phys. Lett. 66B:205 (1977).
6. O. Scholten, F. Iachello, and A. Arima, Interacting boson model of collective nuclear states III, preprint KVI-126, February 1978 and references therein.
7. D. R. Bes and R. A. Broglia, Phys. Rev. C3:2349 (1971).
8. R. A. Broglia, D. R. Bes, and B. Nilsson, Phys. Lett. 50B:213 (1974).
9. R. A. Broglia and D. R. Bes, Phys. Lett. 69B:129 (1977).
10. It has been shown[8] that $G_{\lambda} \simeq 27/A$ MeV, independent of the multipolarity and for $f_{\lambda}(r)=1$. It is thus not evident that only the lowest multipolarities are important. However, following ref. 4 we restrict the present discussion to $\lambda=0$ and $\lambda=2$.

11. The expansion parameter in the NFT is $1/\Omega$, Ω being the effective pair-degeneracy of the valence single-particle orbitals. In the case of a j-shell, $\Omega=j+\frac{1}{2}$.

12. B. Bayman and A. Lande, Nucl. Phys. 77:1 (1966); T. Kishimoto and T. Tamura, A163:100 (1971).

13. G. Racah, Phys. Rev. 63:367 (1943).

14. It may be instructive to compare the Pauli correcting factors given in equations (6) and (7) of ref. 15 with those appearing, e.g. in (7). The factor $[1-(n_s-1)/\Omega]^{\frac{1}{2}}$ corresponds to the first order term in the expansion of $[(2\Omega+2-2n_s-4n_d)/(2\Omega+2-4n_d)]^{\frac{1}{2}}$ in powers of $1/\Omega$. The term $(1-n_d/\Omega)^{\frac{1}{2}}$, which is absent from eq. (7) of ref. 15, is however contained in the reduced matrix element appearing in eq. (6) of the same reference.

15. A. Arima, T. Otsuka, F. Iachello, and I. Talmi, Phys. Lett. 76B:139 (1978).

16. R. A. Broglia, V. Paar, and D. R. Bes, Phys. Lett. 37B:257 (1971).
    D. R. Bes, R. A. Broglia, and B. Nilsson, Phys. Rep. 16C:1 (1975).

17. D. R. Bes, G. G. Dussel, and H. Sofia, Am. Journal of Phys. 45:191 (1977).

18. O. Civitarese, R. A. Broglia, and D. R. Bes, Phys. Lett. 72B:45 (1977).

19. S. G. Nilsson, Mat. Fys. Medd. Dan. Vid. Selsk. 29:no.16 (1955).

ON THE INTERACTION OF MANY-PHONON MONOPOLE PAIRING VIBRATIONS

P. F. Bortignon

Istituto di Fisica Galileo Galilei
Universita di Padova
Padova, Italy

and

R. A. Broglia

The Niels Bohr Institute
University of Copenhagen
Copenhagen, Denmark

In this note we study the properties of a system of many mono-pole pairing vibration phonons[1] in the framework of the nuclear field theory (NFT) (cf. e.g. ref. 2 and references therein). Exact solutions are obtained for the case of two and three interacting bosons. The results obtained are generalized to the case of an arbitrary large number of bosons.

The system of four fermions moving in two-levels as well as in a set of many single-particle levels and interacting via a pairing force with constant matrix elements was considered in ref. 3. We summarize here the corresponding results. For the case of a single j-shell the exact solutions are analytic functions of the pairing coupling constant G and of the degeneracy $\Omega = j + \frac{1}{2}$ (cf. e.g. ref. 4). In particular the interaction energy (anharmonicity) between the two phonons is given by

$$\Delta W = 2G \tag{1}$$

while the two-particle transfer "cross section" from the two to the four fermion system is given by

$$\sigma = \left| <j^4(0)|T|j^2(0)> \right|^2 = 2(1 - \frac{1}{\Omega}) \tag{2}$$

The NFT contributions to ΔW and σ were calculated in ref. 3 up to third order in the expansion parameter $1/\Omega$. The corresponding diagrams displayed in fig. 1 of ref. 3 were worked out in the framework of the Rayleigh-Schrödinger (RS) perturbation theory in terms of the quantities

$\varepsilon = \varepsilon_o + G$;   Hartree-Fock energy

ε_o ;   distance between the two levels

$W_o = \varepsilon - G\Omega$;   phonon energy

$\Lambda = G\sqrt{\Omega}$;   particle-vibration coupling strength.

These quantities completely define the fermion and boson fields.

The lowest order $(1/\Omega)$ contribution (graphs 1a and 1(h) - 1(k) of ref. 3) already give the values (1) and (2). An exact cancellation is found for the contributions of higher order in $1/\Omega$. In particular for diagrams of order $1/\Omega^2$ and $1/\Omega^3$ where the direct and wavefunction normalization contributions are worked out explicitly in ref. 3.

As examples of many-level systems, the ground states of ^{204}Pb and ^{212}Pb are calculated in ref. 3. In zeroth order these states can be viewed as two-phonon pairing vibrational states, i.e.

$$|I> \equiv |gs(^{204}Pb)> = |gs(^{206}Pb) \otimes gs(^{206}Pb)>$$

$$|II> \equiv |gs(^{212}Pb)> = |gs(^{210}Pb) \otimes gs(^{210}Pb)>. \tag{3}$$

The exact results for the interaction energy of the two systems and for the matrix elements of the two-particle transfer operator were obtained dioganalizing the pairing interaction in a basis of four holes (particles) moving around the N = 126 closed shell. The levels included in the calculation were taken from experiment. The correlation energies of the one-phonon states are 640 keV for the ground state of ^{206}Pb and 1237 keV for the ground state of ^{212}Pb. They are reproduced by using for the pairing strengths G the values 0.13 MeV and 0.10 MeV respectively. The resulting values of the interaction energies for the two states are

$$\Delta W_I = 540 \text{ keV}, \quad \Delta W_{II} = 267 \text{ keV}. \tag{4}$$

Note that ΔW_I is of the same order of magnitude as the correlation energy of the one-phonon state.

The NFT results are displayed in table 1. For both the case of ^{212}Pb and ^{204}Pb the $(1/\Omega)$ contribution accounts for 90% or more of the exact value. In both cases, the $(1/\Omega)^2$ and $(1/\Omega)^3$ contributions (which are zero for the two-level model) arise from cancellations between large values. The same kind of agreement between the exact and the NFT results (up to order $(1/\Omega)$) is found for the two-particle transfer operator (see table 1b).

Table 1

$\Delta W(keV)$

Order	I	II
$1/\Omega$	474	262
$1/\Omega^2$	42	4.3
$1/\Omega^3$	14	0.4
	530	266.7
Exact	540	267

(a)

	Exact	NFT
R_1	1.797	1.724
R_2	1.909	1.908

(b)

In (a) are reported the results of the diagonalization and of the NFT for the interaction energies associated with the two-pair addition (gs(^{212}Pb)) and two-pair removal (gs(^{204}Pb)) phonons.
In (b) we display the ratios

$$R_1 = \frac{\sigma(gs(^{206}Pb)) \to gs(^{204}Pb))}{\sigma(gs(^{208}Pb)) \to gs(^{206}Pb))}$$

and

$$R_2 = \frac{\sigma(gs(^{210}Pb)) \to gs(^{212}Pb))}{\sigma(gs(^{208}Pb)) \to gs(^{210}Pb))}$$

calculated utilizing the results of the diagonalization and the $1/\Omega$ NFT diagrams.

We take up next the problem of the three-phonon system for the case in which the fermions are moving in two single-particle levels.

The exact result for the interaction energy among the three phonons is, in this case[4]

$$\Delta W = 6G. \tag{5}$$

The NFT diagrams contributing to ΔW up to order $(1/\Omega)^2$ are displayed in fig. 1. Again, the $(1/\Omega)$ term (graph (a) of fig. 1) already gives the exact result (5). The value of the $(1/\Omega)^2$ terms (graphs (b) – (e) of fig. 1) are:

$$
\begin{aligned}
\text{graph 1b} &= 60\,\frac{G}{\Omega} \\
\text{graph 1c} &= -12\,\frac{G}{\Omega} \\
\text{graph 1d} &= -24\,\frac{G}{\Omega} \\
\text{graph 1e} &= -24\,\frac{G}{\Omega}
\end{aligned}
\tag{6}
$$

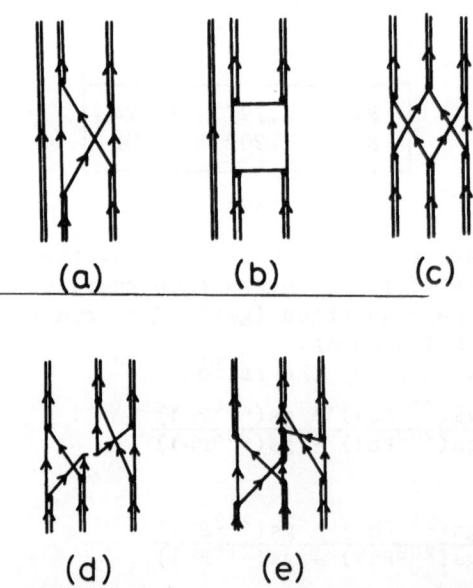

Fig. 1.   The graphs 1a – 1e are the NFT contribution (in RS perturbation theory) up to order $(1/\Omega)^2$ to the phonon-phonon interaction.   The rectangle in 1b represents the diagrams
1b – 1g of fig. 1 in ref. 3.

The "three-phonon" diagrams cancel exactly the very large contribution of the term (1b). Therefore, they are essential to generate the right perturbation expansion.

The NFT contributions up to order $1/\Omega$ give also the exact result for the two-particle transfer cross section between two- and three-phonons states which is equal to

$$\left| <n = 3|T|n = 2> \right|^2 = 3(1 - \frac{2}{\Omega}). \qquad (7)$$

Generally, the $1/\Omega$ contributions to the different matrix elements between monopole pairing vibrations interacting through a pairing force provide an excellent approximation to the exact solution.

To the extent that the nucleus can be described as a gas of s and d bosons (monopole and quadrupole pairing vibrations) it is likely that a single quadratic effective phonon-phonon interaction[5] can account for the anharmonicities of the system. We have shown that such ansatz is correct for s-bosons. It is however an open question whether it is also correct for the case of d-bosons.

REFERENCES

1. D. R. Bes and R. A. Broglia, Nucl.Phys. 80:289 (1966);
    A. Bohr and B. R. Mottelson, Nuclear Structure, Vol. II,
       Benjamin, New York (1975)
2. P. F. Bortignon, R. A. Broglia, D. R. Bes, and R. Liotta,
       Phys.Rep. 30C:305 (1977).
3. P. F. Bortignon, R. A. Broglia, and D. R. Bes,
       Phys.Lett. 76B:153 (1978).
4. B. R. Mottelson, Cours de l'Ecole d'Eté de Physique Théorique
       de les Houches (1958).
5. D. Janssen, R. V. Jolos, and F. Dönau, Nucl.Phys. A224:93 (1974);
    A. Arima and F. Iachello, Phys.Rev.Lett. 40:385 (1978).

ON THE RELATION BETWEEN THE O(6) LIMIT OF THE INTERACTING BOSON MODEL AND TRIAXIAL NUCLEAR MODELS

J. Meyer-ter-Vehn

Institut für Kernphysik
KFA Jülich
D-5170 Jülich, West Germany

It is shown that the O(6) limit of the interacting boson model (IBA) corresponds to the γ-unstable model of Jean and Wilets, if one considers an infinite number of bosons in IBA (N→∞). The rigid triaxial rotor model with γ = 30° satisfies the same selection rules, but differs considerably with respect to B(E2)-values and energies.

1. INTRODUCTION

The IBA model of Arima and Iachello[1] has been very successful in describing low-lying collective states throughout the nuclear table. The model is based on a simple Hamiltonian satisfying SU(6) symmetry. Spectra and transition probabilities have been worked out by systematic application of group theory. The subgroups SU(5), SU(3), and O(6) are of particular interest. It has been shown that, for N→∞, the pure SU(5) limit[2] corresponds to the harmonic vibrator model, and the SU(3) limit[3] to the model of axially symmetric rotors. For the O(6) limit[4], however, it has not been quite clear to which geometrical model it corresponds to[5].

In this contribution, the close relationship between the O(6) limit of IBA and the γ-unstable model of Jean and Wilets[6] is outlined. The essential point is that the γ-unstable model is based on the group O(5) which is contained in the group O(6). The two models yield identical B(E2) ratios for N→∞. On the other hand, it is pointed out that the rigid rotor model[7] with γ = 30° behaves differently.

In the following, we briefly write down the basic equations of the IBA-O(6) model, the γ-unstable model, and the triaxial rotor model with $\gamma = 30^{\circ}$, and discuss their relationship. Some B(E2) ratios are compared explicitly.

2. THE IBA-O(6) MODEL

IBA is based on 5 quadrupole bosons $d_{2\mu}$ plus 1 monopole boson s which generate the group SU(6), and the O(6) limit arises from the group chain SU(6)⊃O(6)⊃O(5)⊃O(3). The IBA Hamiltonian conserves boson number. Therefore one deals with finite representations $[N]$ of SU(6), in general. The O(6) limit[4] of the Hamiltonian can be written

$$H = A\, P_6 + B\, C_5 + C\, C_3 \tag{1}$$

in terms of the quadratic Casimir operators P_6, C_5, C_3 of the sub-groups O(6), O(5), and O(3), respectively. The eigenstates $|[N]\sigma\tau\nu_\Delta LM\rangle$ are labeled by the quantum numbers $\sigma = N, N-2, \ldots 0$ (or 1), $\tau = \sigma, \sigma-1, \ldots 0$, $L = 2\lambda, 2\lambda-2, 2\lambda-3, \ldots \lambda$ where $\lambda = \tau - 3\nu_\Delta \geq 0$ and $\nu_\Delta = 0, 1, \ldots \leq |\tau/3|$. The spectrum has the closed form

$$E([N]\sigma\tau\nu_\Delta LM) = \frac{A}{4}(N-\sigma)(N+\sigma+4) + \frac{B}{6}\tau(\tau+3) + CL(L+1) \tag{2}$$

For $N\to\infty$, only the $N = \sigma$ part of the spectrum, corresponding to a representation of O(6), remains at finite energies

$$E(\tau\nu_\Delta LM) = \frac{B}{6}\tau(\tau+3) + CL(L+1) \tag{3}$$

The O(6) quadrupole operator reduces to

$$Q_{2\mu} = \alpha_2(d_{2\mu}^{+}\sqrt{N-n_d} + \sqrt{N-n_d+1}\,(-)^{\mu}d_{2-\mu}) \Rightarrow \tilde{\alpha}_2(d_{2\mu}^{+}+(-)^{\mu}d_{2-\mu}) \tag{4}$$

Here, $\alpha_2(\tilde{\alpha}_2)$ are adjustable parameters.

3. THE γ-UNSTABLE MODEL

The γ-unstable model of Jean and Wilets represents a special case of Bohr's model of nuclear surface oscillations[8]. Bohr's Hamiltonian

$$H = \frac{D}{2}\sum_{\mu}|\dot{\alpha}_{2\mu}|^2 + V((\alpha\alpha)_o, (\alpha\alpha\alpha)_o) \tag{5}$$

is given in terms of the 5 quadrupole modes $\alpha_{2\mu}$ ($-2 \leq \mu \leq +2$) which generate the group SU(5). The Jean-Wilets model is essentially the O(5) limit corresponding to the group chain SU(5)⊃O(5)⊃O(3). Its geometrical meaning becomes clear after transforming $\alpha_{2\mu}$ to intrinsic coordinates $(\beta, \gamma, \phi, \Theta, \Psi)$

$$\alpha_{2\mu} = \sum_{\nu} D_{\mu\nu}^{(2)}(\phi, \Theta, \Psi) a_{2\nu} \tag{6}$$

with $a_{20} = \beta\cos\gamma$, $a_{22} = a_{2-2} = (\beta\sin\gamma)/\sqrt{2}$, and $a_{21} = a_{2-1} = 0$. Then, the potential energy $V(\beta,\gamma)$ is a function of the deformation parameter β and the asymmetry parameter γ alone. The Jean-Wilets model arises when choosing $V(\beta,\gamma)$ to be independent of γ and having a sharp minimum in β direction. In the pure γ-unstable case, only $(\gamma,\phi,\Theta,\Psi)$ are kept as dynamical variables, but $\beta = \beta_0$ is fixed. The Hamiltonian (setting $V(\beta_0) = 0$)

$$H = \frac{\hbar^2}{2D\beta_0^2} \left[\sum_{k=1}^{3} \frac{I_k^2}{4\sin^2(\gamma - \frac{2\pi}{3} k)} - \frac{1}{\sin^3\gamma} \frac{\partial}{\partial\gamma} \sin^3\gamma \frac{\partial}{\partial\gamma} \right] = \frac{6\hbar^2}{2D\beta_0^2} C_5 \quad (6)$$

then turns out to be proportional to the quadratic Casimir operator C_5 of the group O(5) as shown by Rakavy . The eigenstates $|\tau,\nu_\Delta,LM\rangle$ can be labeled by the Rakavy seniority $\lambda = 0,1,\ldots$ and $L = 2\lambda$, $2\lambda-2$, $2\lambda-3,\ldots\lambda$ where $\lambda = \tau-3\nu_\Delta \geq 0$ and $\nu_\Delta = 0,1,\ldots \leq |\tau/3|$. The energy spectrum is

$$E(\tau,\nu_\Delta,LM) = \frac{\hbar^2}{2D\beta_0^2} \tau(\tau+3) \quad (7)$$

and the quadrupole operator has the form

$$Q_{2\mu} = \frac{Q_0}{\beta} \alpha_{2\mu} = \frac{Q_0}{\beta} \cdot \frac{1}{\sqrt{2}} (n_{2\mu}^+ + (-)^\mu n_{2\mu}) \quad (8)$$

where $Q_0 = 3R_0^2 Z \beta/\sqrt{5\pi}$ is the intrinsic quadrupole moment and

$$n_{2\mu}^+ = \frac{1}{\sqrt{2}} (\alpha_{2\mu} - (-)^\mu \frac{\partial}{\partial\alpha_{2-\mu}}) \quad (9)$$

$$(-)^\mu n_{2-\mu} = \frac{1}{\sqrt{2}} (\alpha_{2\mu} + (-)^\mu \frac{\partial}{\partial\alpha_{2-\mu}})$$

are boson creation and annihilation operators, respectively. In addition, it should be noted that Bohr's Hamiltonian (5) does not conserve the number of bosons, $[H,N_B] \neq 0$ with $N_B = \sum_\mu n_{2\mu}^+ n_{2\mu}$, in general. This is in contrast to the IBA Hamiltonian.

Comparing the Hamiltonian of the γ-unstable model (6) and that of the IBA-O(6) model (1), the equivalence is apparent when choosing a special representation of O(6) (e.g. $\sigma=N$) and $C = 0$. In this case, IBA and the O(5)-Jean-Wilets model have identical spectra. For $N\rightarrow\infty$, also the quadrupole operators (8) and (4) are the same and lead to identical E2-transition probabilities for both models. This is even true for $C \neq 0$ since the Casimir operator $C_3 = L^2$ in eq. (1) is the squared angular momentum and does not change the wave functions.

Wave functions of the O(5)-Jean-Wilets-model have been calculated for a couple of low-lying states by Bès[10]. Using these wave
functions, some of the B(E2)-values have been calculated and are
given in Fig. 1. The transitions follow the selection rule $\Delta\tau = \pm 1$.
The $2_2^+ \rightarrow 0_1^+$ and other transitions are therefore forbidden. Also all
spectroscopic quadrupole moments are zero. More systematic deriva-
tions of the wave functions are found in refs. 11,12. Closed
expressions for B(E2)-values of certain band have been derived in
the O(6) scheme by Arima and Iachello. Specializing the results of
ref. 4, one obtains for ground-band transitions

$$B(E2; (L+2)_g \rightarrow L_g)/B_{20} = \frac{5}{2} \cdot \frac{L+2}{L+5} \qquad (10a)$$

and for ground-band $\rightarrow \gamma$-band transitions

$$B(E2; L_\gamma \rightarrow L_g)/B_{20} = \frac{5(2L+2)}{(L+5)(2L-1)} \qquad (10b)$$

where $B_{20} = B(E2; 2_1^+ \rightarrow 0_1^+) = Q_0^2/(16\pi)$. More such B(E2) ratios can be
obtained from the results given in ref. 5.

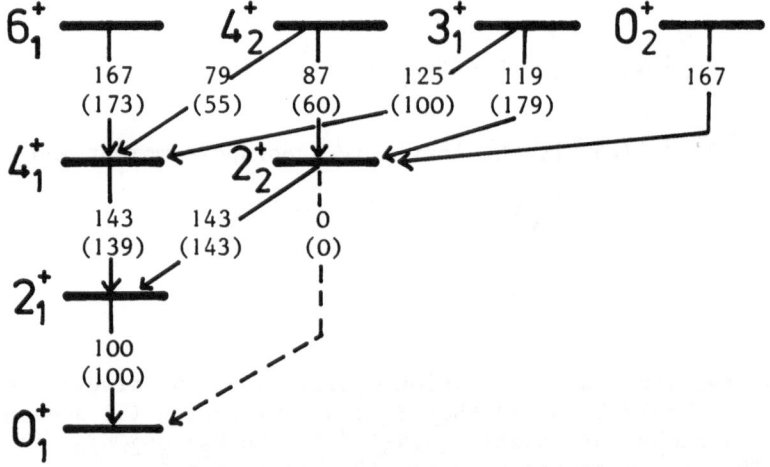

Fig. 1. Relative B(E2) values for the O(5)-Jean-Wilets-model and
the $\gamma = 30°$ triaxial rotor model (in brackets). The 0_2^+
level does not exist in the rotor model. Energy spacings
do not give the actual model spacings. The E2 transitions
$(4_2^+, 3_1^+, 0_2^+ \rightarrow 2_1^+)$ are forbidden due to selection rules in both
models. Also, the quadrupole moments of all states are
zero in both cases.

4. THE TRIAXIAL ROTOR MODEL WITH $\gamma = 30^{\circ}$

The triaxial rotor model is obtained from Bohr's Hamiltonian by fixing γ in addition to β. For $\gamma = 30^{\circ}$, the Hamiltonian (6) then reduces to[13]

$$H = a \left[I_1^2 + 4(I_2^2 + I_3^2) \right] \tag{11}$$

with $a = \hbar^2/(8D\beta_o^2)$. In this case, the wave functions are

$$|LM,\alpha\rangle = \left(\frac{2L+1}{16\pi^2(1+\delta_{\alpha,o})} \right)^{1/2} (D_{M\alpha}^{(L)}(\Omega) + (-)^L D_{M-\alpha}^{(L)}(\Omega)) \tag{12}$$

where $\alpha = 0, 2, \ldots$ L is the projection of angular momentum on the 1-axis, and $D_{M\alpha}^{(L)}(\Omega)$ denotes the usual rotation functions. The energy spectrum

$$E_{L,n} = a \left[L(L+4) + 3n(L-n) \right] \tag{13}$$

is conveniently given in terms of the wobbling quantum number $n = L - \alpha$. For example, the ground-band $(0_1^+, 2_1^+, 4_1^+ \ldots)$ has n=0, the γ-band $(2_2^+, 4_2^+, 6_2^+ \ldots)$ n=2. The B(E2) values are given by

$$B(E2; L_1\alpha_1 \rightarrow L_2\alpha_2) = \frac{5}{16\pi} \cdot \frac{Q_o^2}{2} \cdot \frac{(2L_2+1)}{(1+\delta_{\alpha_1,o})(1+\delta_{\alpha_2,o})} \tag{14}$$

$$\left| \begin{pmatrix} L_1 & L_2 & 2 \\ \alpha_1 & -\alpha_2 & 2 \end{pmatrix} + \begin{pmatrix} L_1 & L_2 & 2 \\ \alpha_1 & -\alpha_2 & -2 \end{pmatrix} + (-)^{L_1} \begin{pmatrix} L_1 & L_2 & 2 \\ -\alpha_1 & -\alpha_2 & 2 \end{pmatrix} \right|^2$$

Apparently, one has the selection rule $\Delta\alpha = \pm 2$ which is equivalent to the $\Delta\tau = \pm 1$ rule in the O(5)-Jean-Wilets case. From expression (13), one obtains the ratios

$$B(E2; (L+2)_g \rightarrow L_g)/B_{20} = \frac{5}{2} \cdot \frac{2L+1}{2L+5} \cdot (1+\delta_{L,o}) \tag{15a}$$

$$B(E2; L_\gamma \rightarrow L_g)/B_{20} = \frac{15}{(L+1)(2L+3)} (1+\delta_{L,2}) \tag{15b}$$

which should be compared with the ratios (10a,b). B(E2) ratios according to formula (14) are also shown in Fig. 1. Although the selection rules for the O(5)-Jean-Wilets case and the $\gamma = 30^{\circ}$ rotor case are the same, the B(E2) values of the γ-band $\rightarrow$ ground-band transitions go $\sim 1/L$ in the Jean-Wilets-model, but $\sim 1/L^2$ in the rigid rotor model. This reflects the difference between "soft" and "rigid" wave functions.

5. CONCLUSION

It has been shown that the O(6) limit of IBA corresponds to the γ-unstable model of Jean and Wilets in the same sense as the SU(5) limit to the harmonic vibrator model and the SU(3) limit to the axial rotor model with degenerate β- and γ-bands. Identical results occur if one considers an infinite number of bosons in IBA. IBA is more general in as far as it allows for representations with finite boson number and corresponding cutoffs.

REFERENCES

1.   A. Arima and F. Iachello, Phys.Rev.Lett. 35:1069 (1975).
2.   A. Arima and F. Iachello, Ann.Phys. (N.Y.) 99:253 (1976).
3.   A. Arima and F. Iachello, Ann.Phys. (N.Y.) 111:201 (1978).
4.   A. Arima and F. Iachello, Phys.Rev.Lett. 40·385 (1978).
5.   A. Arima and F. Iachello, Ann.Phys. (N.Y.) to be published.
6.   L. Wilets and M. Jean, Phys.Rev. 102:788 (1956).
7.   A.S. Davydov and G.F. Filippov, Nucl.Phys. 8:237 (1958).
8.   A. Bohr, Mat.Fys.Medd.Dan.Vid.Selsk. 26, no. 14 (1952);
     A. Bohr and B.R. Mottelson, "Nuclear Structure", Vol. II, W.A. Benjamin, Reading, Mass. (1975).
9.   G. Rakavy, Nucl.Phys. 4:289 (1957).
10. D.R. Bès, Nucl.Phys. 10:373 (1959).
11. E. Chacon, M. Moshinsky, and R.T. Sharp, J.Math.Phys. 17:668 (1976), and O. Castanos, E. Chacon, A. Frank and M. Moshinsky, to be published in J.Math.Phys.
12. A.A. Raduta, V. Ceausescu, and A. Georghe, Nucl.Phys. 311:118 (1978).
13. J. Meyer-ter-Vehn, Nucl.Phys. A249:111 (1975).

IBM VERSUS QUADRUPOLE PHONONS

Vladimir Paar

Prirodoslovno-matematički fakultet
University of Zagreb and "Rudjer Bošković"
Institute, Zagreb, Yugoslavia

ABSTRACT

The main topic of this Symposium is the SU(6) inter-acting-boson model (IBM) of Iachello and Arima. Our aim is to stress its relation to the SU(6) truncated quadrupole-phonon model (TQM) of Janssen, Jolos, and Dönau, as well as to the boson-expansion model and to the Bohr-Mottelson Hamiltonian. The IBM and the TQM are equivalent as far as the phenomenological treatment of the parameters is concerned; among other features, all limiting symmetries (SU(3), SU(5), O(5),...) and the results obtained by fitting the parameters are exactly the same in both models. In fact, the IBM and the TQM are two representations of the SU(6) group (the Schwinger and the Holstein-Primakoff representations, respective-ly). In addition, advantages and shortcomings of SU(6) models versus boson-expansion methods are pointed out. It is noted that the SU(6) Hamiltonian could be inter-preted as a particular realization of the Bohr-Mottelson Hamiltonian.

THE SU(6) TQM OF JANSSEN, JOLOS, DÖNAU - THE ENFORCED CLOSURE OF SU(6) ALGEBRA

While the IBM is well known, the readers may not be familiar with the fact that simple quadrupole phonons also lead to SU(6) symmetry. The SU(6) Hamiltonian for quadrupole phonons has been derived microscopically by Janssen, Jolos, and Dönau.[1-2] It is a quadratic ex-pression in terms of 35 generators of the SU(6) group

$$b_{2\mu}^{\dagger} (N-\hat{N})^{1/2}, \quad (N-\hat{N})^{1/2} b_{2\mu}, \quad b_{2\mu}^{\dagger} b_{2\nu} \quad . \tag{1}$$

Here $b_2^{\dagger}$ is the creation operator of the quadrupole phonon* and $\hat{N}$ is the phonon number operator $N=\sum b_{2\mu}^{\dagger} b_{2\mu}$. The Hamiltonian, hereafter referred to as the μ truncated-quadrupole-phonon-model (TQM) Hamiltonian, has the form:

$$H_{TQM} = H_0 + h_1 \hat{N} + h_2 [(b_2^{\dagger} b_2^{\dagger})_0 (N_{max} - \hat{N})^{1/2} (N_{max} - 1 - \hat{N})^{1/2} + H.C.]$$

$$+ h_3 [(b_2^{\dagger} b_2^{\dagger} b_2)_0 (N_{max} - N)^{1/2} + H.C.]$$

$$+ \sum_{L=0,2,4} h_{4L} [(b_2^{\dagger} b_2^{\dagger})_L (b_2 b_2)_L]_0 \quad . \tag{2}$$

This Hamiltonian was derived in Refs. 1-2 starting from the microscopic Hamiltonian in the quasiparticle representation. It was shown that the commutators of collective fermion operators $(A_{2M}^{\dagger})$ and their combinations lead, in an approximate way, to 35 linearly independent operators, on which the Lie algebra of the SU(6) group is based. The basic approximations in this enforced closure of the SU(6) algebra are as follows: (i) Restriction to the states of angular momentum 2. (ii) Exclusion of noncollective states in structure constants (i.e., in the expressions obtained for commutators). This will be referred to as the SU(6) collective approximation. The corresponding SU(6) collective Hamiltonian, therefore, must be a quadratic expression in terms of these 35 operators. The boson representation expressing the collective fermion operators $(A_{2M}^{\dagger}, A_{2M})$ in terms of phonon operators $(b_{2\mu}^{\dagger}, b_{2\mu})$ and satisfying the same SU(6) commutation relations is of the so-called Holstein-Primakoff type.[3] The transformation $(A_{2M}^{\dagger}, A_{2M}) \rightarrow (b_{2\mu}^{\dagger}, b_{2\mu})$ contains a positive integer, denoted here as N_{max}, which restricts the maximum number of phonons in the wave function. This integer is the eigenvalue of the Casimir operator of the SU(6) group and characterizes the totally symmetric rep-

---

* Here we use the following quantum-mechanical terminology: The phonon operators $b_{2\mu}^{\dagger}, b_{2\nu}$ satisfy boson commutation relations and the phonon space is, in principle, infinite. The collective fermion operator $A_{2M}^{\dagger}$ creates a coherent two-quasiparticle state of angular momentum 2. The state $A_{2M}^{\dagger} |0\rangle_{BCS}$ will be referred to as a collective fermion state.

resentation under consideration. Janssen, Jolos, and Dönau suggested that N_{max} should be equal to half the number of valence-shell particles.[2] However, the microscopic expression for this cutoff value was also derived.[2] It should be stressed that this truncation of the phonon space is not an approximation, but is a consequence of the symmetry properties of the collective fermion operators.

The microscopic expressions for the parameters H_0, h_1, h_2, h_3, h_{4L}, and N_{max} in the SU(6) Hamiltonian (2) are given in Ref. 1. The SU(6) Hamiltonian (2) is transformed[1,2] into the form of Bohr-Mottelson collective Hamiltonian.

THE SU(6) TQM OF HOLZWARTH, JANSSEN, JOLOS - THE ENFORCED SU(6) HAMILTONIAN MATRIX

In the preceding section we have discussed one way of obtaining the SU(6) Hamiltonian for quadrupole phonons. Now we discuss another way starting from the matrix elements. The approach of Ref. 4 starts from the collective fermion space, with the basis states built from collective fermion states, i.e., of the type $(A_2^{\dagger})_{vJ}^N |0\rangle_{BCS} \equiv |NvJ\rangle$. Then, the matrix elements of the microscopic Hamiltonian in this collective fermion space are mapped onto the matrix elements of the corresponding collective Hamiltonian H_{HJJ} in the phonon space, i.e.:

$$\langle NvJ | H | N'v'J' \rangle \approx (NvJ | H_{HJJ} | N'v'J').\qquad(3)$$

Here $|NvJ)$ denotes the N-phonon state of angular momentum J and seniority (and eventual additional quantum number) v. The basic approximations in deriving (3) are analogous to the collective approximation (i) from the preceding section: (i´) The projector is inserted into the fermion matrix elements in (3) and then all non-collective intermediate states are excluded; of the complete set of intermediate states, only those are retained which are built from collective fermion states with angular momentum 2. In this way, the one-to-one correspondence is established between the matrix elements of the microscopic Hamiltonian H in the collective fermion space and the matrix elements of the collective Hamiltonian H_{HJJ} in the phonon space. In this mapping, phonons are used only to describe the coupling matrix elements; at no point is the replacement $A^{\dagger} \to b^{\dagger}, \ldots$ made. The collective Hamiltonian H_{HJJ} (Eq. 11 in Ref. 4) contains norms of the collective fermion states

$$N_{NvJ} \equiv \langle NvJ | NvJ \rangle \quad . \tag{4}$$

The fact that the basis states $|NvJ\rangle$ are not ortho-
normal is a consequence of the Pauli principle between
quasiparticles in the internal structure of collective
fermion states. Therefore, the norms appearing in H_{HJJ}
reflect the Pauli principle. The problem is now reduced
to the calculation of fermion normalization constants
(4). In the approximation of the type (i´), all norms
can be expressed in terms of the norms N_{200}, N_{222},
N_{224}. Using the additional approximation that these
three norms are equal to each other, we obtain the SU(6)
model, i.e., in this approximation

$$H_{HJJ} = H_{TQM} \quad . \tag{5}$$

The physical meaning of the cutoff of the phonon
space is now clear: It reflects the well-defined and
natural inclusion of the Pauli principle. We could say
that the SU(6) model is an approximate way to take into
account the fermion norms for phonon states.

EQUIVALENCE BETWEEN THE IBM AND TQM

The IBM SU(6) Hamiltonian H_{IBM}, expressed in terms
of s and d bosons, is given by Eq. 2.1 of Ref. 5. Considering the matrix elements of H_{IBM}, the authors of
Refs. 5 and 6 have shown that the s degree can be completely eliminated. The parameter N_{max}, which presents
the total number of s and d bosons in the IBM, now
enters the Hamiltonian. In this way, an equivalent
Hamiltonian, namely the TQM SU(6) Hamiltonian (2), is
obtained. Of course, the notation $d^{\dagger}_{2\mu}$ is used instead
of $b^{\dagger}_{2\mu}$. Thus, the IBM Hamiltonian H_{IBM} and the TQM
Hamiltonian (2) are equivalent to the following simple
relations between the parameters:

$$h_1 = \varepsilon_d - \varepsilon_s + \frac{u_2}{\sqrt{5}}(N_{max}-1), \quad h_2 = \frac{\tilde{v}_o}{2\sqrt{5}},$$

$$h_3 = \frac{\tilde{v}_2}{\sqrt{10}}, \quad h_{4L} = \frac{c_L}{2} - \frac{u_2}{\sqrt{5}} \quad . \tag{6}$$

If we treat the SU(6) Hamiltonians H_{IBM} and H_{TQM} phenomenologically (i.e., by fitting or freely choosing
parameters), then the IBM and the TQM give exactly the
same results. The results of numerical fitting, the
limiting symmetries (SU(3), SU(5), O(5),...), the vibra-
tion-rotation transition, etc., are the same in both
models. For example, the SU(3) and SU(5) limits of the

IBM from Refs. 5, 7 are exactly the same as the SU(3) and SU(5) limits obtained in Ref. 8 for quadrupole phonons in the TQM. Therefore, it is not justified to treat these features as an exclusive piece of evidence for the IBM. As long as the parameters are not calculated microscopically, any result obtained in the IBM can equally well be interpreted as a piece of evidence for the quadrupole phonons of the TQM.

In fact, the IBM Hamiltonian, H_{IBM}, and the TQM Hamiltonian, H_{TQM}, are related to two representations of the SU(6) group, i.e., the Schwinger[9] and the Holstein-Primakoff[3] representations, respectively. Let us illustrate these two representations for the SU(2) group. The Holstein-Primakoff representation of the SU(2) algebra in terms of bosons is given by

$$J_+ = B^\dagger (2J-B^\dagger B)^{1/2}, \quad J_- = (2J-B^\dagger B)^{1/2} B, \quad J_0 = -J + B^\dagger B. \tag{8}$$

Here $B^\dagger$ and B satisfy the boson commutation relations, and J is the total spin. In the Schwinger representation, an <u>additional</u> monopole boson $(A^\dagger)$ is introduced, so the spin operators are linear combinations of two kinds of bosons, $(A^\dagger, A)$ and $(B^\dagger, B)$:

$$J_+ = B^\dagger A, \quad J_- = A^\dagger B, \quad J_0 = \frac{1}{2}(B^\dagger B - A^\dagger A) \ . \tag{9}$$

This is only a new representation of the same thing. Using this representation, Schwinger rederived many formulas of Racah algebra.[9] Obviously, the additional monopole boson of the Schwinger representation is simply related to the single boson of the Holstein-Primakoff representation:

$$A^\dagger \rightarrow (2J-B^\dagger B)^{1/2} \ . \tag{10}$$

This can be generalized to the SU(n) group. The IBM and the TQM are examples for n=6. The relation between the operators is

$$\text{IBM} \leftrightarrow \text{TQM}, \quad d_{2\mu}^\dagger \leftrightarrow b_{2\mu}^\dagger, \quad s^\dagger \leftrightarrow (N_{max} - \sum_\mu b_{2\mu}^\dagger b_{2\mu})^{1/2} . \tag{11}$$

What is the difference between the IBM and the TQM? If the parameters are treated phenomenologically, there is no difference between the two models. The models, however, offer different methods for calculating the parameters microscopically. The IBM starts from the interaction between particles, and the TQM starts from the microscopic Hamiltonian in the quasiparticle representa-

tion. However, the calculations have so far been performed mostly phenomenologically. Therefore, they give
evidence neither for the IBM nor for the TQM. It would
be interesting to compare the parameters calculated
microscopically in the IBM and in the TQM for the same
nucleus and for the same microscopic Hamiltonian (as
far as it is possible in the framework of the approximations involved). The question, unsettled as yet, is how
these microscopic parameters compare with each other
and with the phenomenological parameters obtained by
fitting (which, of course, are the same both in the IBM
and in the TQM).

The main approximation of the SU(6) TQM, the
exclusion of noncollective states in the intermediate
states, casts some doubt on the degree of reliability
of the microscopic derivation of the parameters in the
SU(6) Hamiltonian. In fact, it may be possible to
include part of the effect of the excluded noncollective
states into the renormalization of the parameters of the
SU(6) Hamiltonian. This would cause a pronounced
superiority of the fitting procedure over the microscopic calculation.

The microscopic calculation may be even less
reliable in the rotational limit. In this case a self-
consistent feedback would be expected in the intrinsic
structure of the TQM phonon and the IBM bosons.This is
ignored in the SU(6) models.

How can we relate the SU(3) limit of the SU(6)
Hamiltonian to the geometrical picture of deformed
nuclei? The answer has already been given in the TQM.
Since in these considerations the parameters are treated
phenomenologically, the same answer applies to the IBM.
As is well known, any boson system can be transcribed
to the classical picture with the aid of a coherent
state. Such a state permits us to replace the boson
operators by c numbers.[8] Such a situation appears in the
SU(3) limit of the SU(6) Hamiltonian. In this case the
nuclear system can be described in terms of the parameters characteristic of the macrosystem, i.e., nuclear
deformation. Ref. 8 shows that in this sense the collective states in the SU(3) limit present a projection
of the coherent state of the SU(6) group onto the state
of a given angular momentum and its projection. The
SU(3) limit thus leads to the well-known Bohr-Mottelson
geometrical picture of deformed nuclei, and does not
represent a basically new physics.

The IBM is sometimes discussed in terms of proton- and neutron-bosons. It should be stressed that an analogous discussion is also possible in the TQM. In fact, quadrupole vibrations for protons and neutrons separately were considered in Ref. 10. These two quadrupole modes are strongly coupled by the symmetry term in the mass formula. Proton and neutron vibra- tions, thus obtained, are in phase and in antiphase. Vibrations in phase correspond to low-lying quadrupole vibrations.

WHAT IS BETTER: SU(6) MODELS OR BOSON EXPANSIONS (BE)?

By expanding the square roots in the TQM SU(6) Hamiltonian (4), we obtain an SU(6) Hamiltonaan that is an infinite expansion in the phonon operators $b_2^\dagger$, b_2. The SU(6) Hamiltonian, therefore, corresponds to an infinite boson expansion. The usual boson expansions are, in practice, truncated at the fourth- or sixth- order terms in the phonon operators.[*] Therefore, one is tempted to conclude that SU(6) generally gives better results than boson-expansion methods. However, in ad- dition to this advantage, the SU(6) model also incor- porates a serious disadvantage in comparison with the boson expansion. The parameters of the SU(6) Hamiltonian are calculated approximately by neglecting noncollective intermediate states, while in the boson expansion[12] this approximation is <u>not</u> made. Therefore, the SU(6) Hamiltonian corresponds only to an <u>approximate</u> summing of the Marumori boson expansion to <u>infinite order</u>.[**]

---

[*] However, the convergence of the boson expansion is significantly improved if its formulation is given in terms of collective operators.[11] The Marumori expan- sion was modified in this way in Ref. 12.

[**] It should be stressed that the appearance of the square root is not peculiar to SU(6); it arises for any algebra as long as the boson expansion is of the "perturbative type", i.e., a Taylor series in some formal expansion parameter, such as $1/(2j+1)$. This was discussed in Ref. 13 for the Beliaev-Zelevinsky expansion for the SO(2n) case. It does not seem that this implies the validity (exact or approximate) of SU(6) symmetry.

Expansion of the square roots in the SU(6)
Hamiltonian gives rise to anharmonicities up to infinite
order, whereas the BE methods include only terms up to
the (maximum) sixth order. However, already the third-
order terms of the expanded SU(6) Hamiltonian contain
inaccuracies, while the third-order terms of the
(modified Marumori) BE are correct. Thus, in general,
one can not state that the SU(6) model is superior to
BE. This point was illustrated by Holzwarth, Janssen,
and Jolos.[4] They have shown that the truncation of the
boson expansion leads to inaccuracies comparable with
those caused by the exclusion of noncollective states
in the SU(6) model. In the BE of Ref. 12, the effect of
the Pauli principle is included in the matrix elements,
while in the SU(6) model an analogous effect leads to
the phonon cutoff which we have already discussed.

We omit here the discussion of the relation to
other models which in their basis include, in addition
to collective states, also some selected noncollective
states. Such a model is the cluster-vibration model
(CVM), in which the collective states are quadrupole
phonons and the selected noncollective states are the
selected clusters of shell-model particles in the
valence-shell. The CVM provides a description of odd and
even nuclei on an equal footing, and offers an insight
into the specific mechanism for generating a quasi-
vibrational and a quasirotational pattern.[14] In the
cases when clusters are "immersed" into the coherent
structure, the states which appear will approximately
correspond to the collective states discussed earlier.
Otherwise, additional states (causing, for example,
band crossing) appear, without the corresponding collective counterpart.

The author thanks Fritz Dönau, Gottfrid Holzwarth,
Dietmar Janssen, and Eugene R. Marshalek for very useful discussions, and Franco Iachello and Igal Talmi for
the opportunity of expressing these views at this
Symposium.

REFERENCES

1. R.V. Jolos, F. Dönau, and D. Janssen, Construction of
   Collective Hamiltonian in the Microscopic Model of
   Nucleus, Teor.Mat.Fys. 20:112 (1974).
2. D. Janssen, R.V. Jolos, and F. Dönau, An Algebraic
   Treatment of the Nuclear Quadrupole Degree of

Freedom, Nucl.Phys. A224:93 (1974).

3. T. Holstein and H. Primakoff, Field Dependence of the Intrinsic Magnetization of a Ferromagnet, Phys. Rev. 58:1098 (1940).

4. G. Holzwarth, D. Janssen, and R.V. Jolos, On the Validity of the Boson Method for Transitional Nuclei, Nucl.Phys. A261:1 (1976).

5. A. Arima and F. Iachello, Interacting Boson Model of Collective States I. The Vibrational Limit, Ann. Phys. 99:253 (1976).

6. R. Jolos, F. Dönau, and D. Janssen, Construction of Collective Hamiltonian in Microscopic Nuclear Model II, Preprint P4-8077, JINR, Dubna, 1974.

7. A. Arima and F. Iachello, Interacting Boson Model of Collective Nuclear States II. The Rotational Limit, Ann.Phys. 111:201 (1978).

8. R.V.Jolos, F. Dönau, and D. Janssen, Symmetry Properties of Collective States of Deformed Nuclei, Yad.Fiz. 22:965 (1975).

9. J. Schwinger, On Angular Momentum, in: "The Quantum Theory of Angular Momentum", L.C. Biedenharn and H. Van Dam, eds., Academic Press, New York (1962).

10. A. Faessler, E2-Oberflächenresonanzen in sphärischen Kernen, Nucl. Phys. 85:653 (1966).

11. S.C. Pang, A. Klein, and R.M. Dreizler, Study of Boson Expansion Methods in an Exactly Soluble Two-Level Shell Model, Ann.Phys. 49:477 (1968).

12. S.G. Lie and G. Holzwarth, Application of the Boson-Expansion Method to Even Se and Ru Isotopes, Phys.Rev. C12:1035 (1975).

13. E.R. Marshalek, Perturbative Boson Expansions to All Orders for Even and Odd Nuclei, Nucl.Phys. A224:221 (1974).

14. V. Paar, Ward-Like Identities, Cluster-Vibration Model and Quasirotational Pattern, in: "Selected Topics in Nuclear Structure" Vol. 2, J. Styczen and R. Kulessa, eds., Institute of Nuclear Physics, Jagellonian University, Krakow (1978).

GENERAL REMARKS REGARDING THE INTERACTING BOSON MODEL AND

APPROXIMATION

Herman Feshbach

Laboratory for Nuclear Science and Dept. of Physics
Massachusetts Institute of Technology
Cambridge, Mass. 02139, U.S.A.

The object of many of the reports presented at this Workshop has been the understanding of the structure underlying the
observed properties of nuclei in various states characterized
by their excitation energy, their angular momentum, parity and
electromagnetic moments. Adding to these importantly are the
interlevel transitions, particularly those proceeding via the
electromagnetic interaction. These studies have formed a primary
research area of nuclear physics.

The reason for the strong interest in these investigations
stems from the unique character of nuclear matter and of nuclei.
Nuclei consist of relatively few particles, fermions, which interact relatively strongly, although the forces are not as strong as
those which play an essential role in many of the elementary particle interactions. However, except for the very close mutual
encounters, the nucleons move with non-relativistic velocities.
One of the questions, which motivate the study of nuclear states,
and therefore, of central interest for this Workshop, is whether
such a system can have simple modes of motion. We shall want to
discuss this question in more detail below. However, we can give
the answer now and it is yes. This unanticipated discovery is
one of the great triumphs of nuclear research, adding in a profound fashion to our understanding of the relationship between
the forces and states of many body fermion systems. It is remarkable that this understanding was achieved in spite of the fact
that many features of the nuclear forces were not understood at
the time and indeed some remain unclear to this date.

It was found that there are collective forms of motion.

Perhaps the most striking are the shell model and the optical
model which demonstrate that the nucleons generate an average
field in which each of them move. This average field can be calculated using a form of the self-consistent field method. It acts
in place of a strong external central field which in the atom is
the defining field for the single particle orbitals. Moreover,
as is clear from its origin, the shell model field is a dynamic
quantity changing with the nature of the occupied orbitals. The
discovery of the rotational nuclei, of the giant resonances such
as the giant dipole, of isobar analog states shows that.there are
indeed several varieties of specially simple modes of motion which
can be understood in terms of the dynamic shell model. However,
it would appear to be highly likely that there are many more of
these special states; perhaps their signature is as yet not as
highly obvious as those mentioned above. The problem which therefore poses itself asks whether there are new simple modes of
motion as yet undiscovered, how can they be made more visible,
and how can they be related to the basic dynamic shell model. It
is a possible answer to these questions which was under discussion
these last few days.

The objective can be formulated more specifically in terms
of the shell model. As Talmi has pointed out, the existence of
the shell model does not readily produce a solution of the nuclear
structure problem. Rather, as he described, application of the
shell model leads to another set of questions such as: Of the
10^{15} $J = 4^{+}$ levels in the ^{154}Sm, which ones are interesting? Certainly not all of them! In any event, it will be possible to study
all but a tiny fraction. But then which should be studied?

The suggestion, indicated by the history of nuclear structure
studies, is to search for classes of states which are simply related. This is certainly what occurs in the case of the giant
resonances. For example, the wave function for the giant dipole
resonance is obtained to some degree of approximation by operating
on a parent state with the dipole operator. Similar remarks can
be made with regard to the isobar analog state, the low-lying
levels of the vibrational and rotational nuclei and so on. Clearly, it would be a major discovery to uncover a new group of such
states and map out their behavior as the A and Z of the nucleus
changes.

From the point of view of the nuclear Hamiltonian, such simple
relationships are reflected in its invariance properties. If the
Hamiltonian is invariant against a transformation group, its
eigenstates can be classified into sets which form different representations of that group. Because of the group properties,
eigenstates, which are members of a particular representation,
are related by the analogs of the raising and lowering operators,

familiar from the theory of the harmonic oscillator or from the
properties of the angular momentum operator. It is such a relationship which connects the isobar analog to its parent or a
member of a rotational band to the band head. Probes which can
specifically excite these states through an interaction term can
be constructed from the operators forming the transformation group;
that is, it commutes with the Hamiltonian. It is time to comment
that the invariance and group properties need not be exact. But
if the probe operator does commute with the Hamiltonian even
approximately, there will be states which will be especially easy
to excite using the probe, and there will be others whose excitation will be especially difficult; that is, there will be selection rules. More important, because of the group properties, the
ratio of the probabilities for transitions, to different members
of the representation, induced by such a probe will depend only
upon the nature of the group and will not involve any further detailed specification of the states. In the discussion of this
Workshop, particular emphasis was placed on electromagnetic probes,
as well as on the two nucleon exchange reactions such as the (p,t)
or (t,p) reactions. Generally, not all probes are equally effective; that is, their interaction with nuclei, as accurately expressible in terms of operators associated with the group. Of
course, the study of nuclear structure, in this context, involves
not only the search for states which are simply related, but also
the search for probes which will preferentially excite these particular states, with concomitant selection rules. The point being
made is that the interaction of such probes with nuclei will be
expressible in terms of operators belonging to the group.

It is by no means essential the symmetry (or the associated
invariance) be exact. One need only look at the example found in
elementary particle physics to see that the classification of the
states of elementary particles into sub-sets connected by group
properties is extraordinarily useful and provides important in-
sights into the underlying structure, in this case, the quark
structure, even though the symmetries involved are badly broken.
The reasons for the broken symmetry are very significant and there
is considerable effort going into finding what the operating mechanisms are.

These introductory remarks lead to the questions: Does the
IBM (Interacting Boson Model) represent a discovery of a new symmetry in nuclear structure? If it does, can its phenomenological
parameters be related to those of the shell model? Will the IBA
(Interacting Boson Approximation) provide that connection?

In asking these questions, we have distinguished between the
IBM and IBA. The former is a <u>model</u> which assumes the existence
of the U(6) symmetry generated by the s and d bosons, that is 0^+

and 2^+ bosons, involving various phenomenological constants ε, κ and χ. On the other hand, the IBA should provide the justifi-cation for the IBM through an approximation to the general shell model.

    Obviously, one of the questions which must be answered is: What is the value of the transformation to boson space implied by the IBM? The following diagram was drawn several times:

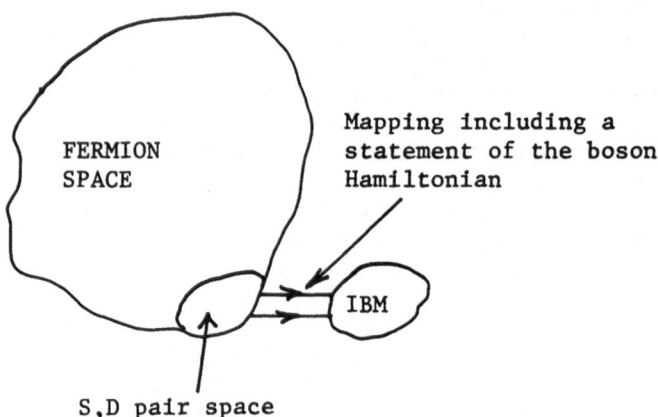

It shows pictorially the transformation to a boson space which is applied to a part of the fermion space, that part in which two nucleons couple in a relative S or relative D state. Why bother to make the transformation? Obviously, if the mapping can be ac-curately stated, then any calculation which can be performed in the boson space can also be performed in the fermion space. And if it is necessary to justify every calculation in the boson space by performing it in the parent fermion space very little will have been gained by the transformation. The value of the transformation lies in that it permits a much more transparent statement of the symmetry under study; more transparent that is than would be pos-sible in the original fermion space. Moreover, and this is not necessarily an independent point, it is an easy matter to diagonal-ize the model Hamiltonian. This is not just a convenience. It is important because it permits the complete determination, for limit-ing cases in analytic terms, of the consequences of the presumed U(6) symmetry.

    The motivation is clear. But now the important question be-comes: Do the nuclei exhibit this symmetry? One can of course attempt to derive it using a-priori considerations. This is the subject of the discussions given by Talmi and Otsuka, Broglia and Bortignon, and also but somewhat less directly by Paar.

Another procedure compares the consequences of U(6) symmetry with experiment. The general pattern of the energy levels, the transition rates, two particle transfer reactions, inelastic electron scattering are of great importance. However, the comparison to prediction has to be done with great care. The symmetry after all is not exact. There are many ways in which it can be broken; for example, the assumption of an inert core is not universally valid. Some of the effects of the core can be taken into account by parameter renormalization. However, states which involve excitation from the core can introduce intruder states to which the model cannot be applied. Even if this is recognized and these levels are not included in the comparison of the experimental data with the model, the interaction of the intruder states with the model levels can perturb the properties of the latter. One must carefully choose the domain in which the model can be applied. One must evaluate the effect of symmetry-breaking mechanisms as examplified by the "intruder" state, determine where they are important and how big they can be. One must know where the model can be wrong and to what degree. To coin a phrase, "If it can't be wrong, it can't be right". We are all familiar with these sorts of consideration since they occur generally when a model is being developed. In the present context, serious deviations from symmetry would be indicated by abnormal behavior of the empirical parameters, κ, ε and χ as the nuclear species is changed.

In addition, there are other models (e.g., the unified model) some of which were presented today. In introducing a radically new model like the IBM, one must establish contacts with the older more established theories. One must determine where do they differ, where are they identical. If they differ, which one is correct? Are there any critical experiments? If the IBM model is to replace older models, it must be breaking new ground and be successful where the older models fail. Some of these relationships have been discussed during this Workshop.

The success of the interacting boson model would indicate a whole new method for looking for symmetries. As an example, are there 4^+ bosons and if so, what kind of symmetry is implied? In addition to the interaction among the 4^+ bosons, it will be necessary to consider the interaction with other bosons, for example, the 0^+ and 2^+ boson. If one restricts one's considerations to the 0^+ and 2^+ subspaces, what kind of broken symmetry is implied by the existence of the 4^+? Comparison with the experimental data would obviously help to determine the impact of the 4^+ boson.

Another type of question asks whether the model can be extended to lighter nuclei? In the mass number range under investigation at present, the particle-particle and hole-hole interaction dominate. Excitation out of the core leads to the possibility of a particle-hole interaction and thus to symmetry breaking and renormalization. The success of the IBM suggests that this last interaction cannot have large effects.

The situation changes when light nuclei are considered. The particle-hole interaction now dominates. In the case of the ^{16}O nuclei, to which the IBM was first applied, Iachello used 1^- and 3^- elementary bosons. The particle-particle and hole-hole interactions were found to be important as perturbations. With the experience obtained by looking at the larger mass numbers, it becomes perhaps profitable to apply the IBM to the light nuclei extending up through Ca.

It has been repeatedly stated that the major area of interest for the IBM are the low-lying levels. It is of course more likely that the corresponding wave functions will have a high degree of symmetry since the attractive potential energy can then be maximized. However, the possibility of relatively highly excited levels with substantial symmetry remains. They may make their appearance as doorway states as exemplified by the giant dipole and the isobar analog. A family of doorway states as a function of Z and/or N would indicate the presence of a symmetry providing an additional motivation for a search for such states.

CONCLUDING REMARKS

Akito Arima

Department of Physics
University of Tokyo
Tokyo, Japan

I would like to begin my concluding talk with a discussion
about relations among several models including the Interacting
Boson Approximation (IBA). As Feshbach pointed out, we should
distinguish the Interacting Boson Model (IBM) from the IBA, because
the IBM is a model while the IBA is an approximation to the nuclear
shell model. The latter was first introduced by Iachello and
Feshbach. We further introduced two different Interacting Boson
Models. One of them assumes one kind of bosons which has two orbits
s and d, while the other does two kinds of bosons with distin-
guishingly proton and neutron. Let me call the first IBM 1 and
the second IBM 2. Some times we call the IBM1 SU(6) model because
s and d bosons altogether belong to the fundamental representation
of SU(6) group.

The IBM 1 has resemblance to the model developed by Jansen
Jolos and Dönau who introduced a cut-off-factor $\sqrt{N-n_d}$ where n_d is
the number operator of d-bosons. One immediately recognizes that
their model can be reduced to the IBM 1 after replacing the cut-
off-factor by s^+ or s, the creation and anihilation operators of
s-bosons.

We assume in the IBM that bosons interact with each other
through a two-body boson-boson interaction. Hamiltonians derived
by boson expansion techniques have interaction terms such as

$$[[d^+d^+]^{(L)}[d^+d^+]^{(L)}]^{(0)} \text{ and } [[d^+d^+]^{(L)}[d^+\tilde{d}]^{(L)}]^{(0)}.$$

These terms are missing in the IBM because the number of s and d bosons is kept constant. In the fixed space coordinate the creation and anihilation operators of d bosons are often expressed in terms of collective coordinates α_μ and the momenta conjugate to α_μ;

$$d^+_\mu = \frac{\sqrt{B\omega}}{\sqrt{2\hbar}} \{(-1)^\mu \alpha_{-\mu} + i\,\pi_\mu/B\omega\}$$

$$\tilde{d}_\mu \equiv (-)^\mu d_{-\mu} = \frac{\sqrt{B\omega}}{\sqrt{2\hbar}} \{(-1)^\mu \alpha_{-\mu} - i\,\pi_\mu/B\omega\}$$

Using them, one can rewrite $[d^+\tilde{d}]$, $[d^+\tilde{d}]^{(0)}[d^+\tilde{d}]^{(0)}$ and other operators; for example

$$[d^+\tilde{d}]^{(0)}/\sqrt{5} = \sum_\mu \{\frac{1}{2B}|\pi_\mu|^2 + \frac{C}{2}|\alpha_\mu|^2\} + \frac{5}{2}\hbar\omega$$

Following Bohr and Mottelson, one can move in the body fixed frame in which

$$d_\mu = \sum D^{(2)}_{\nu\mu}(\theta_i)a_\nu$$

$$a_0 = \beta \cos\gamma$$

$$a_{\pm 1} = 0$$

$$a_{\pm 2} = \frac{1}{\sqrt{2}}\beta \sin\gamma$$

In this frame, a fourth-order interaction $[d^+\tilde{d}]^{(0)}[d^+\tilde{d}]^{(0)}$ can be expressed as

$$[d^+\tilde{d}]^{(0)}[d^+\tilde{d}]^{(0)} = (\frac{1}{2B}\pi^2 + \frac{C}{2}\beta^2 + \frac{5\hbar\omega}{2})(\frac{1}{2B}\pi^2 + \frac{C}{2}\beta^2 + \frac{5}{2}\hbar\omega)$$

$$= \frac{1}{(2B)^2}\pi^4 + \frac{C}{4B}(\pi^2\beta^2 + \beta^2\pi^2) + \frac{C^2}{4}\beta^4 + 5\hbar\omega(\frac{1}{2B}\pi^2 + \frac{C}{2}\beta^2)$$
$$+ (5\hbar\omega/2)^2.$$

Other interactions appearing in the IBM 1 are similarly written in terms of π_μ, β and γ. Namely the IBM 1 has effective mass, π^4 etc β^2, β^4, $\beta^3 \cos 3\gamma$ and thier function. On the other hand in a geometrical picture, its Hamiltonian usually consists of a pure kinetic energy and a boson-boson interaction $V(\beta, \gamma)$. If one goes back to the laboratory coordinate system, the Hamiltonian contains fourth order terms such as

$$[[d^+d^+]^{(L)}[d^+d^+]^{(L)}]^{(0)}$$

$$[[d^+d^+]^{(L)}[d^+\tilde{d}]^{(L)}]^{(0)}$$

and their complex conjugate besides

$$[[d^+d^+]^{(L)}[\tilde{d}\,\tilde{d}]^{(L)}]^{(0)}.$$

The IBM 1 on the other hand contains only the last term, though if necessary the other two terms are simulated in the IBM 1 by

$$[[d^+d^+]^{(L)}[d^+d^+]^{(L)}]^{(0)} \text{ ssss}$$

and

$$[[d^+d^+]^{(L)}[d^+d]^{(L)}]^{(0)} \text{ sss,}$$

They are, however, of higher order and usually ignored in the IBM.

I think that it is relevant here to spend a few minutes to review the ancient history of the SU(6) model. The earliest paper which introduced s bosons was published by Iwamoto (Prog. Theor. Phys. 1958) immediately after Elliott's paper on the SU(3) model. He, however, applied it only to two dimensional problem. Then Arima (1966) and Taruishi (1967) applied the SU(6) model to a description of states of nuclei with several bosons. Those papers treated bosons only macroscopically. They, unfortunately, did not try to analyze experimental data at that time. Jansen, Jolos and Dönau has made a success in arriving at the SU(6) model microscopically in 1974 though having the cut-off-factors instead of s bosons. Iachello applied successfully the Interacting Boson Approximation to medium weight nuclei first using d bosons only and soon introducing s bosons in 1974. This was a very important first step for us to develop our model further. Iachello and Arima in 1974 pointed out that two subgroups of SU(6), SU(5) and SU(3) can be used to describe realistic nuclei. We thus came back to the present time.

Let me proceed to higher order terms which were discussed by Paar, Broglia and Bortignon in this seminar. Paar proved that a similar identity to Ward's identity works well in nuclear physics. He showed for example that an affective M1 operator $[\sigma Y^{(2)}]^{(1)}$ does not contribute very much in vibrational nuclei. His work indicates that there is a mechanism to suppress higher order contributions. Broglia and Bortignon developed their Nuclear Field Theory and showed cancellation among higher order terms. Otsuka's calculation in a large j-shell also showed the similar situation.

A question was often asked during this workshop; "What is a connection between the IBM 1 and the IBM 2 ?" Iachello answered this very clearly. In the IBM 1, SU(6) has four subgroups SU(5), SU(4), SU(3) and SU(2). We always use SU(2). Three sybgroups then remain to describe limitting situations;

$$SU(6) \supset SU(5) \supset SU(2) \qquad \text{anharmonic vibration}$$

$$SU(6) \supset SU(3) \supset SU(2) \qquad \gamma \text{ stable rotation}$$

$$SU(6) \supset SU(4) \equiv O(6) \supset SU(2) \quad \gamma \text{ unstable shape.}$$

According to Scholten's calculation, the IBM 1 and the IBM 2 give almost identical results in the SU(5) and SU(3) regions. In other words, F spin which was introduced by Arima, Otsuka, Iachello and Talmi is reasonably good in those areas. On the other hand in the region of O(6), the F spin seems to break badly. Nevertheless states connected by strong E2 transitions are equally well described and indeed some intimate connection can be mathematically established between the IBM 1 and the IBM 2, though IBM 2 is superior to the IBM 1 concerning energies of some levels and weak E2 transitions. We can thus still use the IBM 1 to systematize data. Now, I would like to discuss "success and failure" of IBM 1. The SU(3) limit gives very good classification of levels in $^{156}_{64}$Gd for example. Scholten showed how successfully the SU(6)(IBM 1) model and the IBM 2 can explain a transitional situation from the SU(5) limit to the SU(3) one taking the Sm isotopes. More precise data are needed to compare the results given by the IBM 1 with those obtained by Kumar, and Kishimoto and Tamura. Casten showed us his beautiful experiment where many branching ratios were observed in the Pt isotopes. One example is the decay scheme of the 1.792 MeV, 4^{+} state which can decay into four lower states: According to the O(6) limit, however, only two transitions are allowed. Indeed the observed branching ratios confirm this selection rule very nicely! The SU(6) model predicts that the ratio of B(E2, L+2→L) to that given by the Bohr-Mottelson model decreases as L increases. Gelberg showed that this is the case in $^{80}_{36}$Kr. On the other hand, this reduction does not occur in Th (as shown by Stephens) and Dy (as shown by Helmling). This feature indicates that some improvement may be needed in the effective E2 operator. More systematical study should be done in near future.

Now, I would like to give comments to a few talks concerning coupling with other degrees of freedom than the quandrupole collectivity. Sujkowski et al. observed a band based on the first 3^{-} state in ^{150}Sm. Paar showed that the Alaga model works well in odd-A nuclei. In this model three nucleons are assumed to form a cluster which in turn couples with even-even core (either rotor or

vibrator). According to Meyer-Ter-Vehn, odd nuclei consist of a
nucleon and γ-deformed core. This model also enjoys a good agreement with the observation. Although some calculations have been
done to describe the structure of odd-nuclei by using the IBM 1,
we need some more calculations by using either the IBM 1 or the
IBM 2. This problem is being investigated by Iachello and Scholten
using the IBM 1 and their results seem to be promissing. Coupling
with two extra nucleons in an intruder orbit (such as $h_{11/2}$ and
$i_{13/2}$) appears to play an important role. (For example Toki and
Faessler and Paar). This effect was also studied by using the
IBM 1 but a further study is strongly desired.

One of the most serious problems in the IBM is its microscopic
foundation. Otsuka discussed a method to truncate a large shell
model space introducing the S-D subspace which turns out to be
mapped onto the sd boson space. He has shown that this truncation
and mapping are good approximation using a large single j shell
$(j=\frac{23}{2})$. However, one needs a further generalization to treat real
nuclei. McGrory supported the basic idea of the interacting boson
approximation (IBA) showing that a main part ($\sim85\%$) of wave functions
indeed belongs to the S-D subspace using the Oak Ridge large shell
model program. An interesting discovery was made by Ginocchio who
took the 0f 1p shell. The largest symmetry group of this shell is
SU(20). He found a Hamiltonian which can be exactly diagonalized
in the S-D subspace which is constructed by two quasi-spins $S^+ S^0 S^-$
and $D^+ D^0$ and D^-. His subgroup chains are

$$SU(20) > Sp(20) \supset SO(5) \supset SO(3)$$

and

$$SU(20) \supset SO(6) \supset SO(3)\ .$$

His mathematical finding is very encouraging and instructive. This
model will play a role such as the Lipkin model to examine any kind
of microscopical models of nuclear collective motion.

Then, what are necessary in the IBA and the IBM? The first is
their microscopic foundation. Feshbach raised a question; what is
an underlying symmetry in s_π, d_π, s_ν and d_ν? We have a picture of
those bosons which was confirmed to be good in systems with a large
single j shell and many j-shells with degeneracy. In many actual
cases one must work with systems with neutrons and protons in many
j-shells which are not degenerate. This is not an easy problem.
From the macroscopic point of view, we must continue to obtain
more data and calculations systematically. In this workshop, we
heard an interesting suggestion from Walters that some 0_2^+ could be

intruders. Besides energies and electromagnetic transitions, aban-
dant information has been provided by the (p,t) reaction and the
(e,e') scattering for example. Concerning the electron inelastic
scattering, Dieprink told us an interesting analysis of
^{150}Nd(e,e')^{150}Nd(2^+) by which he and his collabolators predicted
that a level at 0.850 MeV should be doublets of 2^+ and 1^-.
Kentucky's experiment indeed confirmed this prediction by using the
(n,n',γ) reaction.

I still remember the early stage of many particle nuclear shell
model. One of the best evidences was Talmi's simple mass formula
which was applied to the Ca-isotopes and the N=28 isotones. His
formula has three parameters a, b and ε

$$BE(f_{7/2}^n) = BE(closed\ shell) + a\frac{n(n-1)}{2} + b[\frac{n}{2}] + \varepsilon n.$$

Though no one could calculate those days the values of parameters
from the nucleon-nucleon interaction, this formula worked well.
One thus started to believe many particle shell model calculations.
Even now we have only a qualitative explanation of those parameters
from more foundamental point of view. Because of this education,
I would like to establish first a model and to obtain the values of
parameters phenomenologically. I then proceed later to try to
understand those values microscopically.

Let me finish my concluding talk by emphasizing the importance
of symmetry and simplicity to understand nuclear structure.

PARTICIPANTS

A. ARIMA   Department of Physics,
University of Tokyo
Hongo, TOKYO 113, Japan

P. BORTIGNON   Istituto di Fisica Galileo Galilei
Università di Padova,
I-35100 PADOVA, Italy

R. BROGLIA   The Niels Bohr Institute,
Blegdamsvej 17,
2100 COPENHAGEN ϕ, Denmark

A. BROWN   Department of Physics,
University of Oxford
OXFORD OX1 3NP, Great Britain

R. F. CASTEN   Brookhaven National Laboratory,
UPTON, New York 11973
U. S. A.

M. A. DELAPLANQUE   Institut de Physique Nucléaire,
Université Paris-Sud, 91406
ORSAY, France

A. E. L. DIEPERINK   Kernfysich Versneller Instituut,
University of Groningen,
GRONINGEN, The Netherlands

H. FESHBACH   Laboratory for Nuclear Science and
Department of Physics, M.I.T.
CAMBRIDGE, Mass. 02139, U.S.A.

R. FOUCHER   Institut de Physique Nucléaire
                                        Université Paris-Sud, 91406
                                        ORSAY, France

A. GELBERG   Institut für Kernphysik der
                                        Universität Köln, KOLN
                                        West-Germany

J. N. GINOCCHIO   Los Alamos Scientific Laboratory
                                        University of California,
                                        LOS ALAMOS, New Mexico 87545, U.S.A.

L. HASSELGREN   Institute of Physics,
                                        University of Uppsala,
                                        Box 530, S-75121 UPPSALA
                                        Sweden

H. HEMLING   GSI, Postfach 541
                                        6100 DARMSTADT 1,
                                        West-Germany

K. HEYDE   Laboratorium Voor Kernfysica
                                        Rijksuniversiteit Gent,
                                        B-9000 GENT, Belgium

F. IACHELLO   Kernfysisch Versneller Instituut
                                        University of Groningen,
                                        GRONINGEN, The Netherlands

P. KLEINHEINZ   Institut für Kernphysik
                                        KFA Jülich,
                                        D-5170 JÜLICH, West-Germany

R. D. LAWSON   Physics Division
                                        Argonne National Laboratory,
                                        ARGONNE, Illinois 60439, U.S.A.

M. LEINO   Research Institute for Theoretical
                                        Physics, University of Helsinki,
                                        SF-00170 HELSINKI 17, Finland

G. LOBIANCO   Istituto di Scienze Fisiche,
                                        Università di Milano,
                                        I-20133 MILANO, Italy

N. LO JUDICE   Istituto di Fisica Teorica,
                                        Università di Napoli,
                                        I-80125 NAPOLI, Italy

J. MEYER-TER-VEHN	Institut für Kernphysik, KFA Jülich, D-5170 JÜLICH, West-Germany
J. B. McGRORY	Oak Ridge National Laboratory OAK RIDGE, Tennessee 37830 U.S.A.
A. MÜLLER-ARNKE	Institut für Kernphysik, Technische Höchschule, 61-DARMSTADT, West-Germany
T. OTSUKA	Department of Physics, University of Tokyo, Hongo, TOKYO 113, Japan
V. PAAR	Institute Rudjer Boškovic, ZAGREB, Yugoslavia
G. PUDDU	Kernfysich Versneller Instituut, University of Groningen, GRONINGEN, The Netherlands
R. RICCI	Istituto di Fisica Galileo Galilei, Università di Padova, I-35100 PADOVA, Italy
J. SAU	Institut de Physique Nucléaire, Université de Lyon, LYON, France
O. SCHOLTEN	Kernfysich Versneller Instituut, University of Groningen, GRONINGEN, The Netherlands
F. S. STEPHENS	Lawrence Berkeley Laboratory University of California BERKELEY, California 94720, U.S.A.
Z. SUJKOWSKI	Instytut Badan Jadrowych 05400 SWIERK, near Warsaw Poland
S. SZPIKOWSKI	Institute of Physics M. Curie Sklodowska University PL-20031, LUBLIN, Poland

I. TALMI   The Weizmann Institute of Science,
                                      REHOVOT
                                      Israel

W. B. WALTERS   Chemistry Department,
                                      University of Maryland,
                                      COLLEGE PARK, Maryland 20742,U.S.A.

J. L. WOOD   Chemistry Department,
                                      Georgia Institute of Technology,
                                      ATLANTA, Georgia 30332, U.S.A.

INDEX